Productivity vs Protection

How WinDes Source Control achieves the balance

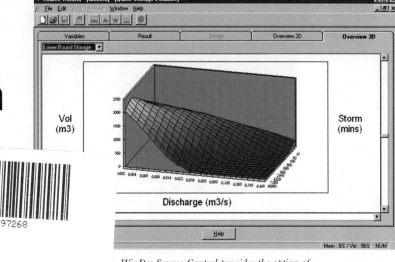

WinDes Source Control provides the option of environmentally friendly alternatives to large storage structures.

The squeeze on the water industry has brought the traditional challenges of engineering into even sharper focus. How do you strike a balance between pressures on productivity arising from environmental concerns, and the downward pressures on costs?

So far, the emphasis has been on CSO discharges and the programme set out under AMP3. However, the next phase will be to look at prevention, rather than cure.

Now the Environment Agency and the Scottish Environment Protection Agency (SEPA) are calling for more environmentally stable storage techniques, bringing the issue of source control firmly into the realm of developers.

Source Control

New technology from Micro Drainage, developers of the industry-standard WinDes drainage design suite, provides the productive power to enable engineers to adopt an innovative approach to the question of source control.

Until now, the use of storage structures has been the usual way of controlling high levels of initial flow,

such as "flash" run-off from paved areas. Too often this solution is unsatisfactory, not only because the structures themselves can be uneconomical, but also because of the associated costs of making them environmentally friendly.

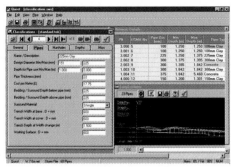

QuOST is a full quantities package, adding more productive power.

The new Source Control module for WinDes enables the engineer to find radical solutions based on the use of porous paving and other infiltration techniques. Source Control provides an estimate of the most economical way of managing the flow and provides swift, accurate solutions where the use of infiltration is appropriate.

With WinDes to perform the laborious and time-consuming tasks (and to provide solutions), much of the workload can be carried by

non-specialists, enabling engineers to concentrate on the most demanding aspects of a project.

Productivity is transformed, while the solutions proposed conform to the latest demands of approving authorities.

QuOST for quantities

A further boost to productivity comes from the QuOST module. QuOST draws data from the other WinDes modules and prepares a full costs and quantities schedule.

Both QuOST and Source Control share the user-friendly and intuitive performance of the existing WinDes suite. Extensive use of graphics makes the software quick and easy to learn and to use, as well as helping to give greater impact and clarity to presentations.

Micro Drainage has always led the way in drainage software. As new challenges emerge to confront the overstretched resources of industry, the new WinDes modules mean that a powerful, productive solution has already been found.

For more information about Source Control, QuOST and WinDes, call Micro Drainage now.

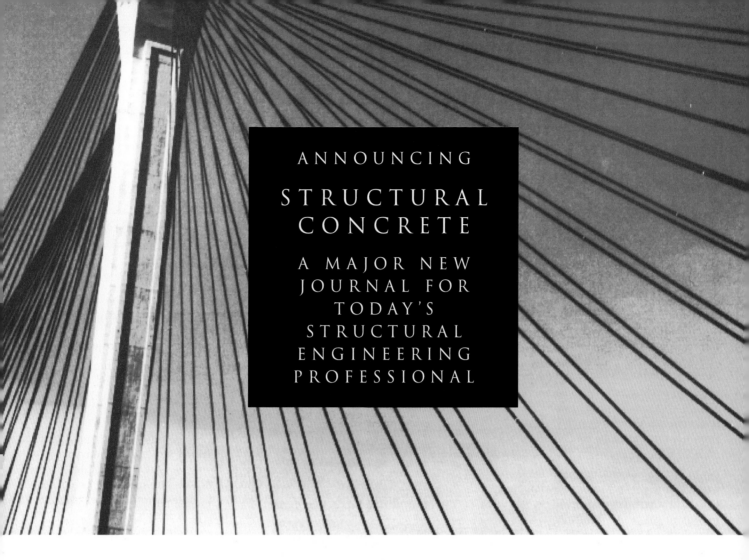
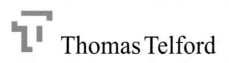

PROCEEDINGS OF
THE INSTITUTION OF
CIVIL ENGINEERS

Millennium Beaches

SUPPLEMENT TO CIVIL ENGINEERING
VOL. 132 • SPECIAL ISSUE 1 • 1999

CONTENTS

Front cover: Civil Engineers have managed an £11 billion programme to clean up the beaches of England and Wales for future generations

Introduction

J. K. Banyard, BSc, AGCI, CEng, FEng, FICE

Welcome to this special issue of *Civil Engineering,* the general journal of *Proceedings of the Institution of Civil Engineers.* It describes the lead role played by civil engineers in one of the world's biggest environmental clean-up operations—the unprecedented £11 billion programme to improve sea water quality around England and Wales.

It is the first of four special issues planned to celebrate the new millennium. Subsequent issues will cover the engineering aspects of sustainable development; the Greenwich Peninsula, showpiece for urban regeneration; and the Jubilee Line Extension, model for future mass transit systems.

When the editorial panel set out to consider suitable topics for the four special issues which would span the end of the 20th century and the beginning of the 21st, initial ideas ranged from historical perspectives to exercises in crystal-ball gazing. Entertaining though these may have been, the panel finally decided that the interests of readers would be best served by focusing on how today's civil engineers are solving today's problems to create a better environment for tomorrow's society.

Awareness of marine pollution is a relatively modern phenomena. At a major ICE co-sponsored conference in 1972 on sewage treatment there was virtually no recognition of the inadequacy of so-called marine disposal—if anything there was tacit support for it. But as the public's concern for the environmental grew, so did the UK's reputation as the 'dirty man of Europe'.

The ensuing programme to raise marine discharge standards along the entire coastline of England and Wales and tackling the extensive legacy of Victorian (and many much later) sanitation systems is the subject of this issue. Despite common problems, the papers reveal there was no common solution which could be repeated at all sites—engineers had to chose from a vast array of complex techniques and adopt them to suit local demands.

The proportion of beaches in England and Wales complying with the minimum standard of the EU's 1975 bathing water directive has risen from 75% to 90% in just 8 years—a tribute to all those involved. But there have also been some disappointments. There are lessons to be learnt in particular about the limitations of modern modelling techniques, which meant results were not always as good as expected. Indeed, the government has recently indicated a requirement for secondary treatment at all sites.

There are also lessons on the need to involve local communities, on the importance of aesthetics and on the potentially negative impact of such developments on the local environment.

My thanks go to all the authors for their time and patience in preparing and revising their papers and to the referees for reviewing them. Please note the papers will be presented and discussed at an ordinary meeting at the Institution in the autumn; please call Lesley Wilson at the ICE on 0171 665 2242 for further details. Written discussion contributions need to be submitted no later than one week after the meeting.

John Banyard, director of asset management at Severn Trent Water, is honorary editor of Civil Engineering and was responsible for the assessment of all papers in this special issue

Bathing waters in the UK— a clear future

E. Thairs, PhD

*Proc. Instn Civ. Engrs
Civ. Engng,
Millennium Beaches*
1999,
132, 3-6

The UK has one of the largest coastlines in Europe—around 18 500 km—and, with no-one living more than 120 km from it, a fascination with the sea second to none (Fig. 1). Not surprisingly, bathing waters are consistently high on the political agenda.

Over the last 30 years policy and practice on bathing waters have changed considerably. The major reasons for change have been European legislation, public perception and expectations—and a £ multi-billion investment programme introduced by the UK water industry. Over 100 major bathing water schemes have been introduced since the EU bathing water directive came into being in 1975, it alone being responsible for some £3000 million capital expenditure. The EU urban waste-waters treatment directive, with a price tag to the water industry in excess of £8000 million, has added to the coastal protection programme against potential sewage pollution.

Recent decisions mean that the water industry will be making even further investments. First, it will provide more widespread secondary treatment around the coast following the UK government's decision to remove high natural dispersion area

(less sensitive area) status. Also it will be accelerating its stormwater overflow improvement programme. These actions should deliver even higher compliance than the 89% achieved now for mandatory bathing waters.

The water industry is also working with the environmental agencies to identify the causes of occasional non-compliance of designated bathing waters, which occur even with major water industry improvement schemes nearby. Bathing water quality can be highly influenced by others—manufacturing industries, agriculture and tourism for instance. Moreover, natural sources of pathogens—from animals and birds—will make absolute compliance impossible in most cases. A partnership approach must be adopted to get to the levels of compliance that the citizens of the UK desire.

EU bathing water directive

The central focus for protecting bathing waters is the EU bathing water directive. This was adopted on 10 December 1975 with 10 years for the standards prescribed for bathing waters to be met, unless derogations were granted. Its purpose was to improve and at least maintain the quality of bathing water so that public health was protected and amenity value secured.

Bathing water is defined as fresh or sea water in

Fig. 1. No-one lives more than 120 km from the UK's 18 500 km coastline

Ted Thairs is an environmental advisor to Water UK

Table 1. Bathing water quality improvements in England and Wales

Year	1987	1988	1990	1992	1994	1996	1997	1998
Number of designated bathing waters	360	364	407	416	418	433	447	457
Number compliant with mandatory minimum standard of EU bathing water directive	224	267	318	328	345	386	397	413
Percent compliant	62·2	75·4	78·1	78·8	82·5	89·1	88·8	90.4

Source: Environment Agency

which bathing was explicitly authorized or was not prohibited and traditionally practised by a large number of bathers. This was open to interpretation. The UK government's interpretation in 1979 was to spell out in an advisory note certain criteria: waters with more than 500 people in the water at any one time; stretches where the number of bathers exceeded 1500 per mile (1·6 km); and stretches with 750–1500 bathers per mile upon agreement between the then regional water authorities and district councils. This led to 27 bathing waters being designated, excluding well-known resorts such as Blackpool. Subsequent advice from authoritative bodies like the Royal Commission on Environmental Pollution (in 1984), a reasoned opinion against the UK by the European Commission, and general public and political pressures, resulted in a further 362 waters being designated in 1987. By 1997, this had risen to a total of 447 and, in 1998, there were 496, including for the first time inland waters.

The directive requires member states to set limits at least as stringent as the imperative (I) and to 'endeavour to observe' the guideline (G) values it spells out for 13 physical, chemical and microbiological parameters. A further six parameters were included, most as secondary indicators of pollution.

The Environment Agency is responsible for monitoring bathing waters in England and Wales and for assessing compliance. There are minimum sampling frequencies—for several parameters fortnightly during the bathing season, 20 times a year; others, such as faecal streptococci and salmonellae, have to be checked when the agency suspects they may be present. To conform with the directive, 95% of samples for parameters where an I value is given must comply with those values, and 90% of samples in other cases. In practice, compliance with I values is assessed for total and faecal coliforms, the main indicators of contamination from sewage, and with G values for total and faecal coliforms and for faecal streptococci.

None of the parameters gives a direct measure of the concentrations of pathogens (including viruses) which might give rise to gastro-intestinal infections, the main cause of public health concern, or infect the eyes, ears, nose or throat. They only serve to indicate sewage in the water and the possibility of risk from this source.

Back in 1959, the Medical Research Council reported that for all practical purposes there was no risk to health from bathing in coastal waters unless the pollution is so gross as to be 'aesthetically revolting'. A study by the WRc for the Department of the Environment in 1994 concluded that the mandatory standards in the directive gave adequate health protection.

Although the European Commission came forward with a proposal in 1994 to amend the directive, this was ill-construed and has not been progressed. Compliance is therefore assessed against the 1975 directive. This has shown progressive improvement, as for example for in England and Wales (Table 1).

In 1998, a record number of coastal bathing waters (349 out of 389 in England) met the directive's coliform standards and, of the 40 which failed, half missed compliance by only 1 in 20 samples.

As the Environment Agency itself acknowledges, most of this improvement is due to the investment made by the UK water industry in over 100 separate schemes, many involving UV disinfection to eliminate microbiological contaminants.

Award schemes

The EU bathing water directive is not the only means by which bathing water and beach quality is classified in the UK. The Tidy Britain Group administers two initiatives—the European Blue Flag scheme and the seaside award scheme. In addition, there is there is now the green coast awards scheme in Wales.

Fig. 2. Bournemouth's Blue Flag beach meets the stringent guideline standard of the EU bathing water directive

Table 2. Implementation programme for EU urban waste-waters treatment directive

Agglomerations	Discharges	Treatment required	Deadline
Over 10 000 pe	To sensitive areas	More than secondary	31/12/1998
Over 15 000 pe	All	Secondary or equivalent	31/12/2000
10–15 000 pe	All	Secondary or equivalent	31/12/2005
2–10 000 pe	To freshwater and estuaries	Secondary or equivalent	31/12/2005
Under 10 000 pe	To coastal waters	Appropriate	31/12/2005
Under 2000 pe	To freshwater and estuaries	Appropriate	31/12/2005

European Blue Flag scheme

The European Blue Flag scheme began in 1989 as an initiative of the Foundation for Environmental Education in Europe. The scheme, sponsored by the European Commission, gives an annual award for high quality recreational beaches. Only bathing waters officially designated under the EU bathing water directive are eligible, and these have to meet guide values for coliforms and faecal streptococci. In 1998, 146 coastal bathing waters in England met the guide value criteria. Other criteria for success include beach cleanliness and management, and safety arrangements (Fig. 2).

Seaside award scheme

The seaside award scheme was introduced in 1992 for the UK. This has two categories

- *seaside award*—water quality must meet the mandatory standards for faecal and total coliforms
- *premier seaside award*—water quality must meet mandatory standards for faecal and total coliforms and guide values for coliforms and faecal streptococci.

Both categories have also to satisfy criteria for litter management and toilet facilities.

Green Coast Awards scheme

The Green Coast Awards scheme is a new initiative being piloted in Wales through the Tidy Wales Group (Fig. 3). It focuses on beaches outside the criteria for Blue Flags, for example bathing waters which are not designated because of their rural location and low public usage.

Other initiatives

The Tidy Britain Group also organizes beach litter surveys. In August 1998 these highlighted concern over the state of beach toilets and the increase in general litter.

Less 'official', but perhaps as influential in raising public awareness of bathing water quality, is the Marine Conservation Society/*Reader's Digest* annual *Good Beach Guide*. The criteria for inclusion draw only vaguely on the bathing water directive, including waters that achieve better than its guide values.

EU urban waste-waters treatment directive

The EU urban waste-waters treatment directive, adopted in 1991, has been the major reason for expenditure on sewage treatment throughout the UK. The directive lays down minimum levels of treatment for collected sewage (Table 2).

Although the directive did allow for less than secondary (biological) treatment for less sensitive areas, the government announced in 1998 the withdrawal of all designations in this category.

Much of the investment—over £8000 million in total—has been or will be made around the coast. Usually, discharges to inland waters had already been provided with secondary treatment. However, the policy for discharges to estuaries and coastal waters was to make use of the environment's natural capacity to break down and dilute sewage effluent. Hence untreated or partially treated sewage effluent was discharged through outfalls, often

Fig. 3. Former Olympic swimmer Sharon Davies promoting the Green Coast Award scheme for high quality rural beaches

long, to points at sea where conditions for degradation and dilution were optimum. This policy ended with the directive. The extra treatment being provided clearly provides additional protection for bathing waters, whether or not they are officially designated.

Other directives

The bathing water and urban waste-waters treatment directives are not the only legislation to drive investments by the water industry which might affect bathing waters. Also notable are the shellfish waters and dangerous substances directives.

Shellfish waters directive

The shellfish waters directive of 1979 requires EU member states to designate waters needing protection or improvement to support shellfish life and growth. It sets values (*I* and *G*) for 12 parameters, notably organohalogens and metals. Originally, the directive was implemented in the UK by way of a guidance note: 27 designations were then made. Regulations are expected in 1999, with a somewhat extended list of waters to be designated.

Dangerous substances directive

The dangerous substances directive of 1976 imposes water quality objectives or uniform emission limits for a range of dangerous chemical substances, notably organohalogens and certain metals. It also requires EU member states to put in place national programmes for regulating other, less dangerous, substances.

Stormwater controls

The armoury of controls over continuous discharges is not by itself sufficient. The bacterial quality of designated bathing waters could be compromised by intermittent stormwater discharges which deposit unhygienic and unsightly solids. To help ensure that this does not happen, the UK water industry has adopted several approaches.

Improved storage arrangements have been implemented so that stormwaters can be held and fed into treatment works when flows have subsided. Several techniques have been developed by the water industry over the last few years, including the use of storage basins and tunnels, permeable pavements and optimizing storage in the sewer network. The costs can be huge: those associated with a 5 km tunnel at Brighton were, for example, in excess of £40 million.

Appropriate design, location and improvement of stormwater systems has also been undertaken, normally in relation to combined sewer overflows (CSOs). These allow excess waters, suitably screened, to bypass the sewage treatment works. There are over 24 000 CSOs in the UK, of which in 1996 up to 30% were considered to be performing at any one time unsatisfactorily from a water quality and/or aesthetic viewpoint. A prioritized programme for improving these unsatisfactory CSOs

over a 10-year period to 2005 is in place, and is a major item of water industry expenditure.

Performance

The UK water industry must ultimately be judged by its performance in meeting legislative standards, satisfying customer and public expectations and helping to improve environmental quality.

As the major discharger in the UK, accounting for some 40% by volume of all discharges, the industry's environmental performance can be judged by the impact it makes on water quality. However, for bathing waters, it is still not the whole solution. Despite all the improvements in recent years

- 10% of designated bathing areas still do not meet mandatory standards and 3% consistently fail
- 63% of bathing waters still fail to meet the more stringent guideline values (1997)
- 4% of sewage treatment works still fail to comply with their consents (1994).

The UK water industry is well aware of what it must do to fulfil its obligations to bathing water quality, and is continuing to do so. But the investments are huge and need time be properly designed and implemented to ensure they have stability and long-term horizons. Above all, the industry must seek to gain public respect as the principal means of protecting the environment and people against sewerage pollution and is not the cause of it (Fig. 4).

Fig. 4. The UK water industry needs to ensure that its bathing water improvement programme is sustainable in the long term

Cleanseas—Wessex Water's coastal clean-up programme

J. G. Jones

Proc. Instn Civ. Engrs
Civ. Engng,
Millennium Beaches
1999,
132, 6-10

Paper 11708

Written discussion closes
15 November 1999

Keywords: Sewage
treatment & disposal;
environment

The paper describes the various elements of the clean-up programme being undertaken on the north and south coastlines of the Wessex Water region to improve the water quality of the region's 43 designated bathing waters. Since 1989, a total of £100 million has been spent on bathing water improvements and a further £325 million is being spent on waste-water treatment projects—including the world's largest membrane filtration plant at Swanage. The investment has resulted in compliance with the mandatory coliform standard of the EU bathing water directive rising from 81–90% in 10 years and remaining at all times above the national average.

The Wessex Water region covers Bristol, Bath, Somerset, most of Dorset and Wiltshire and part of Hampshire, South Gloucestershire and Devon. The resident population of the region is approximately 2·5 million, although this is increased by around 0·3 million due to holiday visitors at any one time during the summer, principally in the coastal resorts.

The economy and commerce of the region are based to a large extent on tourism. Resorts such as Weston-Super-Mare and Bournemouth (Fig. 1) attract many thousands of visitors each year, with their bathing beaches being popular places for leisure and recreation. Public awareness of environmental issues, in particular the microbiological quality of

bathing waters, has significantly increased over the last five years. Pressure groups such as Surfers Against Sewage and publications like the *Reader's Digest*/Marine Conservation Society *Good Beach Guide* continue to maintain this pressure.

The water company recognizes that its activities can impact on the quality of bathing waters and therefore on local tourism and commerce. As such, it is important to deliver the expectations of high quality bathing water, especially in those areas influenced by marine discharges of waste water. To this end the company has developed a long-term policy that delivers sustainable options for the future based on full treatment and appropriate disinfection techniques.

Fig. 1. Bournemouth's Blue Flag beach attracts thousands of visitors each year

Bathing water compliance in the region

In 1998 there were 451 designated bathing waters in England and Wales. Of these, the number that complied with the mandatory coliform standards was 413, that is 90·4%. This showed an improvement of 1·6% on the results from 1997. In the Wessex region, 40 out of 43 bathing waters passed in 1998 (Table 1), giving a compliance rate of 93·0%, continuing the performance of the previous nine years with above-average compliance statistics.

Bathing waters in the region are shown in Fig. 2. Details of the compliance record since the water company was privatized show a relatively consistent performance, but undoubtedly the influence of rainfall impacts on sea quality.

In excess of £100 million has been spent in the region on capital improvements related to bathing water quality between 1989 and 1994. The prime objective of the EU bathing water directive was to achieve full compliance by 1995; that is, within the 'AMP 1' period. To achieve this aim, as well as the improvement work, the water company relied heavily on the use of disinfection by chlorine at several north coast sites.

Gareth Jones is director of environment and quality at Wessex Water Services Limited

Table 1. Wessex Water bathing water compliance with mandatory coliform standards

	1989	1990	1991	1992	1993	1994	1995	1996	1997	1998
Designated bathing waters	38	39	39	39	42	42	42	42	43	43
Waters not complying with directive	7	0	3	3	7	2	2	1	4	3
Compliance	81·6%	100%	92·3%	92·3%	83·3%	95·2%	95·2%	97·6%	90·7%	93·0%
National compliance	75·8%	78%	75·4%	78·8%	79·4%	82·3%	89·2%	89·1%	88·8%	90·4%

As can be seen from Table 1, bathing water compliance peaked in 1990 but, in the subsequent years up to 1994, random bathing water failures occurred. Detailed investigations could not confirm that continuous effluent discharges were the cause of failures, but heavy rainfall events and the influence of riverine inputs could account for some if not all of the failures.

Urban waste-waters treatment directive

The EU urban waste-water treatment directive (91/271/EEC) governs the treatment of waste water arising from coastal populations. This directive, implemented in the UK by the Urban Waste-Water Treatment (England and Wales) Regulations 1994, establishes minimum treatment standards and timetables for their introduction.

A number of coastal towns are affected by the directive and continuous discharges to coastal or estuarine waters requiring schemes to meet these new statutory obligations have been delivered or are nearing completion. The benefits to bathing water quality arise from the fact that these improvements will reduce the bacterial concentrations discharged to sea. Primary treatment by settlement reduces bacterial numbers by 10% and secondary treatment by between 50 and 90% dependent on the treatment process used.

In this second phase of investment, 'AMP 2', the Environment Agency (EA) focused on the need to improve bathing water quality further. To secure more consistent bathing water quality as well as improvements required by the treatment directive, the EA looked towards disinfection of coastal discharges requiring ultra violet (UV) disinfection systems to be installed, replacing chlorine-based disinfection systems used in earlier years.

Concern was emerging about the bio accumulation of chlorinated bi-products in marine species and, in line with the EA's precautionary principles, it wanted to phase out chlorine by the year 2000. The scale of this commitment in the region is seen by reference to Fig. 3. Work at 22 waste-water sites around the coast, costing around £325 million in the ten-year investment period up to 2000, will be completed.

South coast improvements

By the end of 2000 the water company is planning to spend around £63 million at three sites (Fig. 4) along its southern coastline. The work is particularly significant, for this stretch of coastline

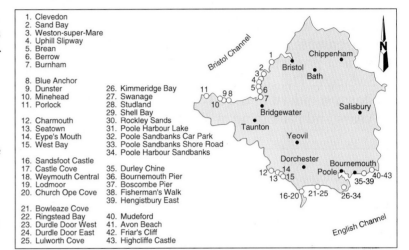

Fig. 2. (above). Wessex Water designated bathing waters

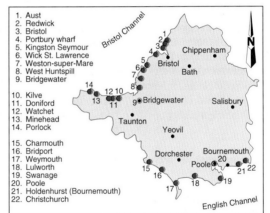

Fig. 3. Coastal sewage treatment works in Wessex Water region

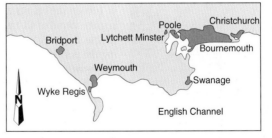

Fig. 4. Locations of south coast improvements

is where tourism, recreation and shell fisheries are very important to the local economy. The sensitivity of many of the locations requires an innovative approach to the design of the works, taking the level of treatment beyond the minimum legal standards, so the landscape or heritage of the area is not adversely affected.

Environmental improvements have already been

completed in a number of areas along the south coast. At Bournemouth, £14 million was spent on the extensions and improvements.

At the Christchurch works, ultra violet treatment has been installed to provide effective disinfection all year round for large quantities of waste-water effluent and to protect local bathing waters. The treatment goes beyond the requirements of present European legislation. Another £2 million is planned to be invested to improve bathing water quality further in Christchurch harbour by improving stormwater arrangements in the catchment.

At Palmersford, £12 million has been spent and, at Poole, £35 million was invested to provide an advanced biological aerated filter (BAF) plant—the largest of its type in Europe (Fig. 5).

In addition, minor improvements to the sewerage system have been carried out at Boscombe, Bournemouth Pier, Hengistbury Head and Poole.

New or modified waste-water treatment works are being built in popular holiday resorts such as Swanage, Weymouth (Wyke Regis) and West Bay (Fig. 6). Additional major improvements are also being made to the works at Christchurch, Bournemouth and Berry Hill.

At Swanage, a combination of membrane filtration and biological treatment is being pioneered. When completed in 2000, Swanage will have the largest membrane waste-water treatment works in the world. The final effluent from this plant will be of a consistently high quality. The introduction of membrane technology in Swanage will provide the town with a level of advanced treatment to take it into the 21st century. The process produces a fully disinfected effluent designed to reduce the environmental impact, of particular importance for a tourist and coastal resort.

North coast improvements
Severn Estuary

The Severn Estuary receives large volumes of run-off from a land mass approaching 20% of the combined areas of England and Wales. The rivers Severn, Wye, Usk, Avon, Parrett and many other minor watercourses contribute significant bacterial inputs to the Estuary.

Water movement in the estuary is complex, but there is a net counter clockwise movement which sweeps water in to the bay at Weston-Super-Mare and then in a northerly direction before turning seawards again in the vicinity of the old Severn Road Bridge. Residence time in the estuary is upward of 200 days, with T90s (the time required to reduce bacterial levels by 90%) approaching 200 hours. Against this backcloth, achievement of guide values required for the awarding of a Blue Flag in reality is challenging.

Schemes on the north coast include the construction of secondary treatment at Bristol waste-water treatment works for 213 Ml/day. Sequencing batch reactors—comprising an area of three football pitches—are in the process of being built, fol-

Fig. 5. (above). Advanced biological aerated filter plant at Poole, largest in Europe

Fig. 6. West Bay beach, site of new waste-water treatment works

lowing the judicial review of the estuarine boundary set by the previous government for the purpose of the waste-water treatment directive.

Although the Severn Estuary is an area of high natural dispersion, derogations are not being progressed and secondary treatment is required. Approximately £50 million will be spent on the project.

Further down the coast, secondary treatment is being installed at Portbury, Kingston Seymour (due in 2000), West Huntspill, Watchet (due in 2000) and Minehead (due late 1999). At most of these sites UV treatment is being provided to deliver year round disinfection.

Weston-Super-Mare

Weston-Super-Mare is a very popular holiday resort. In recent years the local authority has spent a lot of money in improving the quality of services for holiday makers, including greatly improved management of the beach.

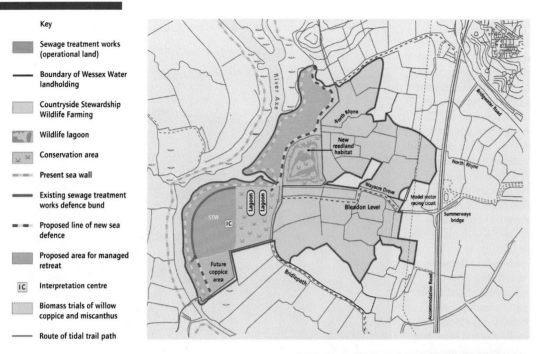

Fig. 7. Site of waste-water treatment works at Bleadon, south of Weston-Super-Mare

Key

■ Sewage treatment works (operational land)

— Boundary of Wessex Water landholding

Countryside Stewardship Wildlife Farming

Wildlife lagoon

Conservation area

Present sea wall

Existing sewage treatment works defence bund

Proposed line of new sea defence

Proposed area for managed retreat

[IC] Interpretation centre

Biomass trials of willow coppice and miscanthus

— Route of tidal trail path

In the 1950s the council embarked upon re-sewering of the town and environs—conveying all the waste water to the southern end of the bay at Black Rock. There they constructed a pumping station which pumped the collected waste water after screening through a short outfall to the Severn Estuary. However, storm outfalls remained in the vicinity of the main beach.

During the 1970s, in an attempt to improve the bacterial quality of the sea, disinfection was practised at the pumping station using chlorine. Improvement in quality was achieved, but performance was patchy; this being a function of the lack of contact time and a wide variation in flow.

Flow variation at Black Rock is vast—up to 24 dwf passing the screens in storm conditions. In an attempt to improve the quality of effluent, a state-of-the-art pumping station was constructed in the 1980s together with improved disinfection of dry weather and storm flows. So great was the improvement in bacterial quality, that in 1990 Weston-Super-Mare was awarded a European Blue Flag.

It had been recognized that these measures were only a short-term solution and, within AMP 2, funding was secured to construct a waste-water treatment works at a cost of around £42 million. The improvements will provide a higher standard of waste-water treatment, reducing the risk of environmental pollution in the area. This work will enable the water company to meet the latest UK and EU regulations.

The waste-water treatment works is being built on former farmland at Bleadon, south of Weston-Super-Mare (Fig. 7). Waste water produced by 100 000 people as well as local industries will receive secondary treatment and UV disinfection. A unique feature of this site is that even storm flows up to 1000 l/s will receive full treatment, including disinfection. Storm storage is effected in a 2·7 km long tunnel, which runs from the existing pumping

Fig. 8. Membrane treatment works at Porlock, first operational plant of its type in Europe

station to the new treatment site.

Completing the north coast Cleanseas programme is the new membrane treatment works at Porlock, set in National Trust land (Fig. 8). The works replaces three crude discharges via short sea outfalls that have caused environmental problems in Porlock Bay for many years. The Porlock works is the first operational plant of its type in Europe. The plant uses advanced membrane technology and was designed in conjunction with the Kubota Corporation of Japan. This process uses ultra-fine filters rather than conventional waste-water treatment to remove the pollutants.

The decision to install the Kubota plant in Porlock was taken after a successful three-year trial of the technology at Kingston Seymour, near Bristol. It was chosen in preference to a conventional waste-water treatment works because of its ability to produce a very high quality effluent in a very confined site. The plant is small enough to be inte-

*Fig. 9. Chesil beach
near Weymouth has
been significantly
improved*

grated easily into the local environment.

As the works are situated in a national park, the water company has worked closely with English Nature, the National Trust and the national parks to ensure that there is as little impact on the surrounding environment as possible. The end result is a disinfected effluent that is 10 times better quality than that required for an EU Blue Flag and more than a million times better than conventional treated waste water.

Conclusions

The Wessex schemes deliver effluents which allow the standards of the bathing water directive to be achieved, and make their contribution to the achievement of Blue Flags.

Customers, hoteliers, local authorities and surfers all want better than minimum standards, not just for the bathing season but all year round. Furthermore, communities which do not have designated bathing waters expect to have equivalent standards. Understandably, they do not want to have inferior standards in comparison to their coastal neighbours. Blue Flags are now commonplace recognition of the quality of a coastal resort and undoubtedly influence the tourist when choosing a resort to visit.

In 1995 the water company, with the assistance of Forum for the Future and Surfers Against Sewage, developed a long-term bathing water policy based on viable and sustainable treatment and disinfection methods which states

'Wessex Water plans that all effluent will receive secondary treatment in line with the requirements of the urban waste-water treatment directive (UWWTD) and, furthermore, discharges that have the potential to

impact on recreational waters or shellfisheries will be disinfected'.

The policy represented a major step forward and was well received in the region. It preceded the recommendations of the House of Commons select committee enquiry into waste-water treatment and disposal. The recent government guidelines to the director general of water services, *Raising the Quality,* clearly recognize the recommendations of the committee and the wishes of customers to have high quality waste-water treatment processes operating all year round.

To be confident that bathing water quality is secure, more attention will have to be paid to the influence of

• combined sewer outfalls
• riverine inputs
• land run-off
• tighter EU standards.

Of these, CSOs could represent significant additional expenditure in future investment periods. In the interim, research needs to be carried on non-chemical disinfection of stormwater discharges. Turbulent flow UV disinfection looks promising, but much research needs to be done to confirm the robustness of this procedure.

An inevitable tightening in European standards will result in the revisiting of a number of completed coastal schemes to add further stages of treatment. What is clear is that significant investment at the coast has delivered much improvement in bathing water quality (Fig. 9), but perhaps in the eyes of customers even more needs to be done.

Southern Water's coastal clean-up programme

S. Peacock, BSc, CEng, MICE, MCIWEM and G. Setterfield, BSc, CEng, FICE, FCI-WEM, MIPR, MIMgt

Proc. Instn Civ. Engrs
Civ. Engng,
Millennium Beaches
1999,
132, 12-19

Paper 11710

Written discussion closes
15 November 1999

Keywords: Sewage
treatment & disposal;
environment; tunnels
& tunneling

This paper describes the programme of works carried out over the past decade around the coast of south-east England by Southern Water. The coastal programme, which has a value in excess of £1 billion, is among the largest of any water company. The paper illustrates the variety of the schemes within the programme by reference to a number of individual projects that serve to illustrate how the particular location for the project has determined the solution adopted. Reference is made to the public relations aspects of the programme and the lessons learned. A schedule of the projects with the scale, designers and main contractors is included.

The area covered by Southern Water includes the counties of Hampshire, West and East Sussex and Kent and the Isle of Wight. The company treats waste water for a total population of about 4·25 million people at 425 waste-water treatment works. Around 65% of the population live in coastal or estuarial towns. The area has 30% of the UK's designated bathing beaches, stretching over 1000 km of coastline from Barton-on-Sea in the west to Broadstairs in the east (Fig. 1).

This paper briefly describes the approach taken by the water company to implement the bathing water regulations and then the urban waste-water regulations over the past two decades. Perhaps the most distinctive feature is the absence of a standard package approach—several schemes utilize long sea outfalls, but equally there are many variants upon even that theme. Some combine treatment for two or more towns; there are traditional works and there are less than traditional approaches (Table 1).

Philosophy

The water company's philosophy is to meet relevant legislation at the lowest cost to customers, while building schemes which blend with the local environment and create the minimum impact on the local community. Inherent in each scheme is the flexibility to accommodate future changes in legislation without excessive expenditure.

The basic approach to managing the programme has been to ensure that each major scheme (or programme of smaller schemes) has been managed by an experienced project manager 100% dedicated to the project. Generally, this will be a member of the water company's staff who has responsibility and accountability for the project.

Comprehensive reporting of the project in respect of programme and cost is put in place to give visibility to all relevant events and activities. Proactive management of both design and construction stages has been the order of the day, with real supply chain management. Visits to suppliers of vital mechanical and electrical equipment have been important in ensuring delivery dates are hit.

The form of procurement is selected on a scheme-by-scheme basis. In some cases partnering

Stephen Peacock is planning and development manager at Southern Water Project Delivery Group

Graham Setterfield was formerly investment director at Southern Water and is now Water UK's director of water services

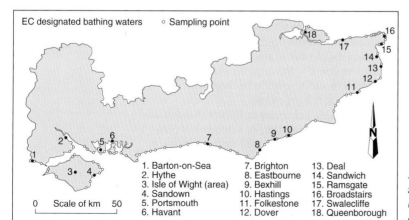

EC designated bathing waters ○ Sampling point

1. Barton-on-Sea	7. Brighton	13. Deal
2. Hythe	8. Eastbourne	14. Sandwich
3. Isle of Wight (area)	9. Bexhill	15. Ramsgate
4. Sandown	10. Hastings	16. Broadstairs
5. Portsmouth	11. Folkestone	17. Swalecliffe
6. Havant	12. Dover	18. Queenborough

0   Scale of km   50

Fig. 1. Southern region showing designated bathing waters and principal towns

Table 1. Summary of Southern Water coastal clean-up schemes

Scheme	Location	Date completed or statutory deadline	Design population	Description of scheme	Designers	Principal contractors
Ashlet Creek	Southampton	2000	15000	Secondary	Mouchel	Babcock
Brighton Tunnel		1998		6m dia. tunnel, p. stn	Acer	Taylor Woodrow
Budds Farm	Havant/Portsmouth	2000	474000	P. stn, tunnel transfer, secondary, tunnel, 5·9 km lso	Halcrow	Nuttalls, Amec
Dover/Folkestone		1999	166000	Tank sewers, p. stns, transfers, primary, 3·3 km lso	S.Water, Mott Macdonald	Otvb, Tilbury D., Millers, Van Oord
Dymchurch/ New Romney		1993	12000	Secondary, uv	S.Water	Nuttalls
Eastbourne		1995	143000	P. stns, transfers, primary, 3·2 km lso	S.Water	Biwater / D&C
Gravesend		2000	62000	Secondary wtw	Montgomery Watson	Birse
Hastings/Bexhill		2000	164000	Tank sewers, p. stns, transfers, primary, 3 km lsos	Not awarded	Millers(part)
Herne Bay		1996	54000	Secondary, return to river	S. Water	Mowlem
Hythe		1998	40000	Primary, 2·8 km lso	Montgomery Watson	Laing
Isle Of Wight		2000	137000	P. stns, transfers, primary, 3·1 km lso	Kaeverner/Brown & Root	lso—Van Oord, wtw — Kaeverner/Brown & Root
Lewes		2000	17000	P. stn, transfer,	Design/build	Docwra (part)
Littlehampton/ Bognor		2000	156500	P. stns, transfers, primary, 3·3 km lso	Halcrow	transfers—Nuttalls wtw—Paterson Candy
Margate		2000	121000	Primary, 1·9 km lso	Not awarded	
Motney Hill	Medway Towns	1999	387000	Secondary wtw	Montgomery Watson	Laing
Newhaven		2000	43000	Primary, 2·6 km lso	Mott Macdonald	Gleeson (part)
Pennington		1996	61000	Secondary	Montgomery Watson	Birse
Portobello	Brighton	2000	319000	Primary, 1·8 km lso	Mott Macdonald	Not awarded
Queenborough		1998	116000	Secondary wtw	Montgomery Watson	Birse
Sandwich Bay	Ramsgate/Deal/ Sandwich	1995	115000	Tank sewers, p. stns, transfers, secondary, return to river	S. Water, Babtie	Gleeson, Christiani-N
Shoreham		1996	63000	Primary, 3·1 km lso	S. Water/Mott MacD.	Gleeson/D&C
Swalecliffe	Whitstable	1998	37000	Primary, 2·6 km lso	Montgomery Watson	Laing
Worthing		1998	139000	P. stns, transfers, primary, 4·5 km lso	S.Water	Gleeson/Alfred McAlpine

Notes
wtw, waste-water treatment works; lso, long sea outfall; P. stn, pumping station.
Excludes requirement for secondary treatment due to withdrawal of HNDA.
Excludes requirements for ultraviolet disinfection due to shellfish directive.

was used, where the nature of the project was of opportunities for innovation and where programmes were demanding. This approach, believed to be the first in the water industry, has allowed a number of 'fast track' schemes to be delivered under budget to critical milestone dates previously thought unachievable. Examples are the urban waste-water projects at Hythe, Swalecliffe and Queenborough in Kent and Morestead in Hampshire.

Diversity of solutions

The varied nature of the sites throughout the region dictates that a large diversity of solutions were chosen to best meet local needs and con-

straints. The following project summaries illustrate this diversity.

Brighton

The town of Brighton, with a population of about 300,000, has been well served by a system of Victorian sewers from which the flows gravitate to a long sea outfall to the east of the town. These Victorian sewers work well under normal flows, but under storm conditions the discharges flow directly into the sea at a number of storm outfalls along the popular tourist beach (Fig. 2).

After extensive investigation, an off-line storm tunnel was selected as the solution. This had the significant advantage of minimizing the disruption to the Brighton seafront during the construction period. The tunnel is 6 m dia. and is 30 m below the beach, and can hold up to 150 000 m³ of storm water. This stormwater is then transferred for treatment to the Portobello works, which serves Brighton, once the flows in the sewers have abated.

The tunnel will be used on average 30 times a year, although it will completely fill infrequently. The tunnel is off-line so tunnel cleaning presented some interesting operational considerations (Fig. 3). The project cost a total of £50 million and was used for the first time in March 1999.

Construction was not straightforward due to a combination of geology and ground water, coupled with slow progress of the tunnelling machine. However, now that it is complete the bathing water at Brighton will be significantly improved. The nearby treatment works is to be rebuilt to meet the standards set under the EU urban waste-waters treatment directive (UWWTD) by the end of the year 2000. Planning permission and consent approval are awaited at the present time.

Combined sewer overflows

The water company's second periodic review obligation is for 126 storm overflows to be improved to meet tighter urban waste-water standards. In most cases, both the discharge frequency is to be reduced and its quality improved. The nature of the work varies considerably; in some instances a manual screen is adequate (where forecast discharge frequency is low; that is, once in five years), in others a major storage provision is necessary (as in the Hastings Stormwater tunnel where 52 000 m³ of tunnel storage deep beneath the town is under construction (Fig. 4)).

Combined sewer outflow (CSO) locations are spread across the region, with clusters on the Isle of Wight and in Hastings and Bexhill; these have been chosen specifically to provide added protection to designated bathing beaches.

For 1997/98 a target of 46 CSOs was agreed with OFWAT as one of the water company's key obligations for that year. In view of the extremely tight programme, it was decided to progress 66 schemes to give some margin for a number not being completed in time. The programme was split in nine separate elements as many of the CSOs fell naturally within the remit of UWWTD schemes currently under construction.

Another important decision was to establish a small team to provide an overview of the whole programme. This had the benefit of giving a clear picture of the overall output prospects and allowed management attention to be focused where it was most needed, by spotting potential bottlenecks and resource shortfalls in time to take effective corrective action.

The Environment Agency (EA) needed to visit all sites to allow them to be officially signed off and thus be verified as outputs to OFWAT. Proactive dialogue was maintained with the EA and this

Fig. 2. (above). Aerial view of Brighton beach showing West Pier

Fig. 3. Cleaning vehicle in Brighton stormwater tunnel

Fig. 4. 'Janice' tunnel boring machine for Hastings stormwater tunnel

resulted in an agreed schedule of visits and a collaborative relationship.

The 1997/98 total achieved was 67, significantly exceeding the target agreed with OFWAT.

Such was the tightness of the programme, that the last 28 outputs were signed off by the EA on 27, 30 and 31 March.

Dover/Folkestone

The project to provide waste-water treatment for the two towns of Dover and Folkestone is now well under way, despite a slow beginning. A planning application for a works serving Folkestone alone was rejected by the planners in 1992; this resulted in a project reappraisal and a scheme combining the two towns was proposed. After extensive consultation, this received approval in 1995.

The scheme, with an estimated project cost of £120 million, includes extensive tunnelling in Folkestone, the construction of two major pumping stations in the town and transfer pipelines and tunnels to the works which are being built on a site between the two towns (Fig. 5). The work is on a site of outstanding natural beauty, and for this reason is being built completely underground. The pumping station at Folkestone Junction, on a former Railtrack site, is impressively large, being 39 m deep and 20 m dia.

The flows from Dover are similarly pumped to the new works where, after treatment, the combined flows are gravitated to Dover and thence out to sea via a long sea outfall into the deep waters of the English Channel (Fig. 6). The work has primary treatment with space left for the addition of biological treatment as legislation now deems. A total of 12 contracts will be awarded during the construction phase of this scheme. The population served is 145 000.

Eastbourne waste-water treatment works

A waste-water treatment works for a population equivalent of 130 000, sited in the beach below sea level, only 30 m from the sea on one side and 50 m from housing on the other, demanded creative thinking from the water company's designers and project managers. The project and its associated long sea outfall cost £42 million by the time it was commissioned in 1996.

Selection of the site appeared straightforward, as the existing pumping station at Langney Point receiving flows from Eastbourne town was hemmed in by residential development to the west and north and by the then proposed Sovereign Harbour development to the east. There were no other viable sites in the vicinity and no rivers where an alternative inland discharge may have been an option. Despite this, there was initially significant local resistance to the scheme, despite the major benefits it would bring to water quality.

In order to have any prospect of success in such a prominent position it was necessary to focus on several issues of major concern to local residents.

Fig. 5. (above). Folkestone transfer tunnels

Fig. 6. Laying storm outfall pipes at Dover

In order of concern these were: odour, traffic and visual appearance. Odour was addressed by adopting chemical scrubbing technology, at the time unusual on the UK water industry. Concerns over traffic movements, particularly sludge tanker movements, were mitigated by introducing sludge centrifuges to allow the use of a smaller number of skip lorry movements.

The most significant feature of the scheme, and the one that facilitated it being constructed in its present location, is the largely underground construction and the design of the superstructure to resemble a 'redoubt', or Napoleonic fort, which is a familiar landmark further along the Eastbourne seafront (Fig. 7). This design was selected from several prepared by the Eastbourne Borough Council's architect's department and for which public views were sought.

The underground box in which the treatment works has been built is a major civil engineering structure. It is 130 m long by 40 m wide, with the eastern third 14 m deep to the base slab, the remaining two thirds 11 m deep. Around 200 tension piles, 43·5 m deep, were required to counteract flotation, together with 37 bearing piles to the same depth which provided the foundation to support the structure during construction. The walls were constructed by the diaphragm wall technique, with bentonite being used to support the excavation. At the peak of construction activity a total of 13 cranes were operating on the restricted site.

The engineering of the scheme is covered in a separate paper. It is very much both a landmark and showpiece project for the water industry, and the work is now regarded by the people of Eastbourne as a major asset.

Portsmouth

The proposed Portsmouth and Havant waste-water treatment scheme will serve 400 000 people and has three main strands. In order to comply with the UWWTD, the discharge from the Eastney long sea outfall into mid-Solent will be enhanced by the introduction of primary and secondary treatment.

In achieving compliance with the UWWTD, the opportunity will be taken to combine the Portsmouth flows with those from the Havant catchment, and to construct a new works serving both communities. The main sites at Eastney on Portsmouth urban area and Budds Farm at the north of Langstone Harbour will be linked by a 7 km long tunnel, 3 m dia., which will take raw waste-water north and return the combined treated effluent to Fort Cumberland from where it will be discharged through the existing 5·8 km long outfall (Fig. 8).

The area abounds with environmental designations, ranging from a site of special scientific interest to a special protection area for wading birds. The tunnel was selected as the transfer medium partly due to these very designations, but also to minimize the otherwise major disruption from the flow transfer construction.

Extensive site investigation has been undertaken, guided by an experienced tunnelling contractor. A local option to construct an underground treatment works at Fort Cumberland (Fig. 9) was considered but set aside due to planning, environmental, archaeological and geotechnical issues.

An important feature of the scheme is the removal of the continuous discharge into the head of Langstone Harbour, resulting in a significant reduction in the organic and micro-biological load

Fig. 7. (above). Superstructure of Eastbourne waste-water treatment works designed as Napoleonic fort

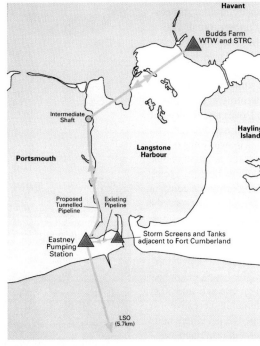

Fig. 8. Plan of Portsmouth and Havant waste-water treatment scheme

in such a highly environmentally designated area.

At Budds Farm, as well as the complete reconstruction of the waste-water treatment works, there will be a new sludge recycling centre utilizing anaerobic digestion followed by thermal drying to produce granules. Maintaining the performance of the existing works is a critical issue during construction, and early involvement of the contractor in discussion with operational staff has been very important in planning the phasing of the work.

A number of planning permissions for the scheme were granted by Hampshire County Council and Portsmouth City Council during 1998, and work is well under way.

Sandwich Bay

The scheme at Sandwich Bay contrasts with most of the solutions adopted as part of Operation Seaclean. The work was completed in 1995 and was opened by HRH The Prince of Wales.

Though conventional, it was far from straightforward. The three towns of Ramsgate, Sandwich and Deal had no treatment; in Deal there was a history of serious flooding which the project was required to solve. The solution has been to collect all normal flows from the three towns and transfer them to an inland site on the River Stour, where they are discharged after full activated sludge treatment.

The discharge into the tidal stretch of the river has the added benefit of improving river-water quality, with the possible longer-term advantage of additional water resource abstraction upstream. A total of 22 000 m^3 of storage has been created by the construction of a series of tunnels under Ramsgate; these flows are then transferred from a new pumping station via pipelines to the works.

The flows from Deal and Sandwich are also transferred via new pipelines; a total of 24 km of pipelines has been constructed. New storm outfalls were built in Ramsgate and Deal by under sea pipe-jacking, the first use of this technique in the water company's area. The population served by the project is 90 000 and the scheme has cost £54 million.

Isle of Wight

The Isle of Wight scheme will collect waste water from the Isle of Wight's coastal towns—in a clockwise order these are Norton, Woodvale and Gurnard, Cowes, Ryde, Bembridge, Sandown, Shanklin and Ventnor—and divert the flows to a new combined waste water treatment works and sludge recycling centre at Sandown, where it will be treated. The treated waste water will then be discharged through a new long sea outfall into the deep waters of the English Channel south east of Sandown Bay.

Improvements will also be made to ensure the frequency of storm discharges around the coast is reduced, and that when spills do occur they are fine-screened to remove unsightly debris. As well as significantly improved water quality in Sandown Bay, a major benefit of the scheme will be that the continuous discharges from the towns on the north coast of the island into the Solent will be removed.

The laying of many km of cross-island mains presents a major challenge for all those involved and will require the co-operation and support of the local community, as will the importation of construction materials for this £100 million project.

Planning permission was granted in late 1997 and construction work is underway.

Herne Bay

Conventional wisdom at Herne Bay in north Kent would have been to install a long sea outfall into the deep water off the coast; however, the shortage of available drinking water in 1989 and 1990 caused

the alternative option of discharge into a local river to be adopted. This increases base flows throughout the year and hence provides a water resource availability in east Kent.

The scheme comprises tunnel sewers connected to a conventional pumping station on the seafront and thence via pipelines to a new activated sludge treatment works. The finally treated effluent is transferred 5 km inland to the River Upper Stour.

Public relations and planning

For each scheme a comprehensive public relations exercise is undertaken, including public meetings, visits to existing operational sites, exhibitions at local venues such as sports centres and libraries. There are many other meetings and discussions with various interest groups, and often a significant amount of correspondence.

Many schemes attract significant media attention and staff regularly appear on local radio and television. Particular attention is paid to concerns over odour and traffic, which together with the quality of bathing waters have proved to be the major topics of interest.

Once on site it is normal practice to establish a liaison group where local people can speak to project staff about concerns they have, and in many cases influence aspects of how the work are carried out. Freephone telephone numbers are provided to encourage a dialogue.

Experience has shown the benefit of advising local people before an event which may cause disturbance rather than dealing with complaints at a

Fig. 9. Aerial view of Fort Cumberland, too sensitive even for underground treatment works

Table 2. Southern Water bathing water compliance by region

	1988	1989	1990	1991	1992	1993	1994	1995	1996	1997	1998
Isle of Wight											
Pass	4	6	10	8	8	12	12	11	12	10	13
Fail	9	7	3	5	5	1	1	2	1	3	0
Hampshire											
Pass	6	10	12	12	11	11	11	11	10	11	13
Fail	5	1	0	0	1	1	1	1	2	1	0
Sussex											
Pass	10	18	14	15	17	18	15	21	22	21	24
Fail	12	4	8	7	5	4	7	1	2	3	1
Kent											
Pass	7	11	12	10	15	17	15	19	18	25	25
Fail	12	8	7	10	5	3	5	1	2	1	1
Total	65	65	66	67	67	67	67	67	69	75	77
Pass	27	45	48	45	51	58	53	62	62	67	75
Fail	38	20	18	22	16	9	14	5	7	8	2
Percent pass	41·5%	69%	73%	67%	76%	87%	79%	93%	90%	89%	97%

Note

Pass or failure is assessed against the following mandatory standards for total and faecal coliforms: faecal coliforms: 2000 per 100ml; total coliforms: 10 000 per 100ml.
A bathing water will comply if 19 out of 20 samples conform. If less than 20 samples are collected in a bathing season, all samples must conform.

Table 3. Southern Water bathing water non compliance 1996–1998

1996	1997	1998	Cause identified/results of investigations	Improvement a result of capital scheme
Folkestone	Folkestone	Folkestone	Stade CSO and/or Copt Point outfall	Improvement in 2001 once Dover/Folkestone scheme operational
Southsea	Southsea		Investigations ongoing	Sealing of sewers will have had some benefit
Hastings	Hastings		Failures attributed to CSO discharges and vandalism	Completion of some K2 dual manholes may have had an effect
	Seagrove		EA/SWS tracer survey revealed a suspected cross-connection between foul and surface water systems. Being investigated	
	Totland Bay		EA/SWS tracer survey revealed a suspected cross-connection between foul and surface water systems. Being investigated	
	Pevensey Bay		Investigations ongoing. Pollution narrowed to one area. Residents report frequent blockages of private sewers	
	Bognor Regis		Unsatisfactory CSOs	Completion of K2 CSO scheme will have improved compliance in wet weather
	Gurnard		Cracked sewer found and successfully repaired (confirmed by tracer exercise). Misconnection to PS outfall removed. (Failures not a result of CSO discharges)	
		Felpham	Thought to have been contributed to by blocked sewers backing-up and leaking into Aldingbourne Rife. Contamination of surface water discharges into Rife also suspected. Investigations ongoing	
Viking Bay			Cracked sewer found and repaired	Will be further improvement from UWWTD scheme in 2001
Newhaven			Results of EA/SWS surveys attributed failure to discharge from Lewes WTW	Will be improvement in 1999 from Lewes transfer to Newhaven
Ryde East			Leaking council sewer contributing to problem. Currently being repaired. CSO discharges also suspected.	K2 CSO scheme ongoing
Stokes Bay			Discharge from council private sewer and contamination of highway drain by caravan club.	

later date. A key feature of keeping the local community informed is the production of scheme brochures and regular newsletters with updates on latest progress.

The approach to planning applications is to agree the terms of reference for site selection with the local authority and to maintain a regular dialogue with them and other key consultees as the feasibility studies and associated environmental assessment work progresses. This process can take up to two years, as many locations have very significant environmental designations and feasible options are limited.

In many cases there is immediate local opposition to schemes, but generally the ground work proves effective and only very occasionally is a scheme refused planning permission.

Environmental benefits

Whereas some environmental pressure groups may argue that the standards of treatment should be higher, there can be no doubting the improvements made to date in the quality of bathing water along the south and south-east coast of England. In 1988 there were only 41% of the beaches meeting the mandatory standard; by 1998 this had risen to 97% (Table 2) with only two beaches failing the mandatory bathing water standard (Table 3).

The benefits have not been limited to bathing water alone. River water flows in east Kent have been improved by the schemes at Herne Bay and Ramsgate. At Brighton a former tip has been transformed by the spoil material from the Brighton tunnel into a park and sports fields (Fig. 10). Many thousands of trees have been planted at a number of sites and some previously sub-standard works have been transformed in appearance and odour removal.

Estuarial water quality will also be significantly improved to the benefit of the many people who use these stretches of water in the south.

Lessons learned

The majority of the lessons learned are not earth-shattering but the continued accumulation of experience in delivering multi-million pound projects. There is no doubt that the more attention to detail that can be given at the feasibility stage, and the better the range of options considered, the more likely there will be savings made against budget.

The involvement of a contractor at the feasibility stage was adopted for the majority of schemes. The process of value management workshops has been integrated into all major schemes after its use on the urban waste-water projects at the early stages.

Another area where lessons have been learned is in respect of the planning process for a programme of this scale. Professional consultants have been an integral part of the teams to provide links with the several county and local councils, the staff of which have at times have been almost swamped with the planning applications associated with the major schemes.

On the public relations front it is clear that however much effort is put in, there will always be additional activities that could have been done. Each scheme has a PR plan which involves identifying the key influencers in the community and keeping them informed and identifying those affected and keeping them involved.

There have been accusations that the company has communicated but not consulted; this remains a hurdle that is difficult to overcome. Dialogue with council officers and members is often done long

before the scheme is public; there will always be some who do not like the location of the works or who would like higher standards of treatment. Regrettably, the chosen solutions are often a compromise between those which different sections of the community favour.

Every scheme will be beset with difficulties; the experience of the programme gives confidence in our ability to overcome these hurdles. The involvement of the operational staff at an early stage has been a lesson well learned. This ensures that there is sufficient familiarity with the chosen solution that commissioning can be well planned from an early stage; it cannot be treated as an afterthought.

Despite the brickbats during the early and construction stages of some of the projects, there is an overwhelming feeling that once complete the finished article is welcomed by the community.

Conclusions

The work that has already been completed and that under construction has, and will have, a profound impact upon the community in the south of England. The quality of bathing water has improved by quantum leaps over the past few years and will continue to improve.

Future legislation may decree that further stages of treatment are necessary; the cost of these and the impact upon the charges will need to be balanced against the benefits that accrue.

The seaside towns of the south were improved by the Victorian generation of public health engineers who are now regarded so highly. It is the belief of the authors that future generations will look back on Elizabethan water engineers in similarly high regard.

Fig. 10. Sheepcote Valley tip restored using Brighton tunnel spoil

Clean Sweep at Penzance/St Ives and Lyme Regis

R. J. Baty, CEng, FICE, FCIWEM, MIMgt, ACIArb, FIWO

*Proc. Instn Civ. Engrs
Civ. Engng,
Millennium Beaches
1999,
132, 20-27*

Paper 11711

*Written discussion closes
15 November 1999*

*Keywords: Sewage
treatment & disposal;
environment; tunnels
& tunneling*

South West Water has spent £900 million since 1989 to ensure that all its 134 designated bathing beaches meet the EU bathing water standard. The company's Clean Sweep coastal clean-up programme covers 33 sites along a coastline of which 70% is subject to environmental protection. This paper focuses on the planning, design and construction of two innovative schemes to provide first-time sewage treatment at Penzance and St Ives in Cornwall and at Lyme Regis in Dorset.

Elimination of the discharge of crude sewage to the sea through outfalls around the coastline of the south-west peninsula of the UK has been a major challenge for South West Water Limited (SWWL) since privatization of the water industry in 1989.

The £900 million capital expenditure involved in achieving this objective comprises a wide range of individual projects embraced under the banner heading of the company's Clean Sweep programme. This was designed to ensure that all of the region's 134 designated bathing waters (out of the national total of some 430) meet the mandatory standard of the EU bathing water directive.

With the addition of treatment standards to other discharges to meet the requirements of the EU urban waste-waters treatment directive (UWWTD) within the next 3–5 years, the 220 crude outfalls positioned around the coastline will have been replaced or upgraded to discharge high quality treated waste water in line with legislation and modern society's expectations. This will be a major achievement on behalf of the region and a credit to all concerned—designers, contractors and local communities.

Clean Sweep itself embraced discharges at 33 geographic locations (Fig. 1). Later schemes, at Sidmouth, Dawlish, Newquay, Camborne and Redruth, Saltash, Ernesettle and Croyde are being promoted to meet the UWWTD requirements, increasing the number of major and interim coastal engineering projects in the region to 44. The benefit of this programme is a monumental environmental transformation of coastal water quality around the region, which many thought would be impossible.

Fig. 1. Clean Sweep coastal clean-up schemes

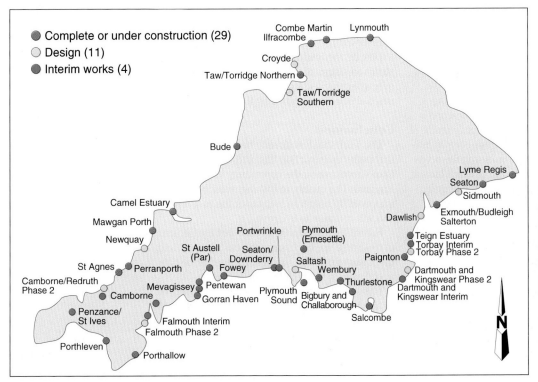

*Robert Baty is chief
executive of South
West Water Limited*

Promotion of individual projects was never likely to be easy given that over 70% of the south-west coastline is subject to understandable environmental protection such as sites of special scientific interest, areas of natural beauty and coastal preservation. Sensitive handling of the promotion, design and implementation of schemes has been a significant ingredient in achieving success. The development of a range of technologies to provide the most environmentally and cost-effective solutions at particular locations has been a key driver as the programme has unfolded.

Solutions vary from traditional sewage treatment through accelerated aeration plants and chemical process plant to simple reed bed technology. At almost half of the locations ultra-violet tertiary treatment is an integral part of the treatment process. This variation in approach to promote cost-effective environmentally sensitive solutions at specific locations can be illustrated by the innovative solutions developed to provide first-time sewage treatment at Penzance and St Ives in Cornwall and at Lyme Regis in Dorset.

Both were major civil and process engineering projects which, to bring to fruition, required a team approach and close working relationships with the communities in areas where livelihoods and well being depend on tourism and maritime activity.

Penzance and St Ives

Located at the south-western tip of the British Isles, the Penzance and St Ives project has been the largest single project carried out within the Clean Sweep programme. At a total cost of £105 million, it has enabled 47 outfalls which discharged raw sewage into Mount's Bay to the south of the peninsula (Fig. 2) and St Ives Bay to the north (Fig. 3) to be abandoned, thus ensuring that eleven identified bathing waters (four in Mount's Bay and seven in St Ives Bay) meet the EU mandatory bathing water standard.

Scheme options evaluated for effluent dispersal and treatment included

- two separate outfalls to serve Penzance and St Ives, with continuation of Hayle STW to serve the existing Hayle catchment only
- single outfall at Lelant on the north coast to serve the region
- single outfall at Hayle to serve the region
- single outfall at Gwithian to serve the region (selected option)
- activated sludge works at Newtown, Penzance, activated sludge works at St Ives, Hayle flows treated at Hayle WWTW
- activated sludge works at Newtown, Penzance, St Ives and Hayle treated at extended Hayle WWTW
- activated sludge works for the whole region at Hayle with existing Hayle WWTW retained for Hayle (selected option).

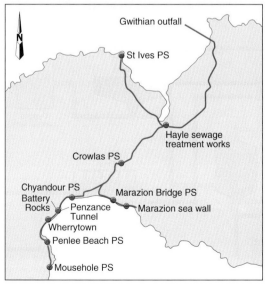

Fig. 2. (top). Marazion beach in Mounts Bay

Fig. 3. (above). St Ives beaches

Fig. 4. Layout of Penzance and St Ives scheme

The reasons for the selection of options adopted for implementation were

- the shallow enclosed nature of Mount's Bay and water movements ruled out dispersion of treated effluent into the bay due to concerns for lack of dilution and eutrophication risks
- construction of separate treatment works for each community (Hayle, Penzance and St Ives) would not have given a lowest whole-life cost solution
- the Hayle WWTW site was capable of development and improvement for the increased flows transferred from Penzance and St Ives
- removal of the treated effluent discharge from the Hayle River, with transfers of all effluent to a deep water discharge off Gwithian, represented the best practical environmental option for both coastlines and their rivers.

Collection of flow from the 47 outfalls and its transfer to a central location for treatment prior to discharge through an outfall constructed in an under-sea tunnel, from which the treated waste water was discharged into the open waters well clear of St Ives Bay, determined the basic layout of the project (Fig. 4).

However, detailed engineering considerations were necessary to manage flows and provide storage capacity within the sewerage network to balance consent-limited combined sewer overflow operation, at times of high rainfall, with the need to provide operationally stable pumping regimes for the many pumping stations on the system. These were required to transfer sewage flows from their former discharge points to an enlarged and upgrad-

Fig. 5. Chyandour pumping station, Penzance

ed centralized sewage treatment works at Hayle.

Collection of flows from the southern Penzance coastline comprised construction of an interceptor sewer 700 mm dia. in Mousehole before pumping through a 200 mm rising main to Penlee progressively increasing in diameter through an 1800 mm dia. rock-bored tunnel from Penlee to Wherrytown West. This leads on to a rock-blasted lined tunnel 3·4 m dia. from Wherrytown West to Chyandour. The tunnel size was designed to provide the storage necessary to balance high flow rates during adverse weather with economic pump capacity.

Pipelines of 250 mm dia. convey flows from Marazion to the east of Chyandour to Hayle WWTW. At Chyandour a major pumping station housing three variable speed and six fixed speed pumps, supported by standby power generation, was constructed on reclaimed land offshore of the original sea wall, which itself was constructed to protect the main line railway link to Penzance (Fig. 5).

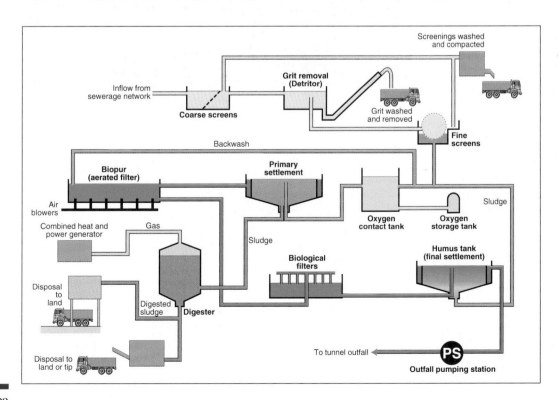

Fig. 6. Process diagram of Hayle wastewater treatment works

From Chyandour on the south coast crude sewage is pumped around 8 km to the treatment works near Hayle. Crude sewage from the north coast is collected in a holding tank at St Ives and pumped via under-sea pipelines across the entrance to St Ives harbour, to the Hayle works from a partially buried and secluded pumping station in St Ives.

The sewage treatment works at Hayle was reconstructed to provide preliminary screening, followed by primary settlement and secondary treatment through a Biopur aerated filter system. Sludge from the treatment process is dealt with by the digestion process with generated gas being used in combined heat and power units to augment power supplies to the treatment works and to provide process heat. The layout is shown diagrammatically in Fig. 6.

Treated waste water from the Hayle works is then conveyed a further 8·4 km to be discharged through a 2·65 km long 900 mm dia. outfall beyond Gwithian on the North Cornwall coast.

The outfall was laid in a 2·8 m x 2·5 m finished profile under-sea tunnel constructed by blasting through Mylor slates (mudstone and shale) for the initial stages, and through Gransthasco Beds (sandstone and siltstone) beyond. This was an innovative cost-saving approach offered by the contractor in preference to a traditionally laid sea-bed outfall. It enabled work to proceed without concern for weather windows and other constraints of traditional marine engineering and proved advantageous to all parties.

At the discharge end of the tunnel, four diffusers at 40 m intervals were drilled and installed through the sea-bed and connected to the outfall pipe laid through the tunnel; flooding of the tunnel itself was then permitted by natural inflow leakage (Fig. 7).

Construction of the project extended over 3 years with work programmes adjusted to meet local needs, in particular ensuring minimal disturbance in the tourist season (Table 1). The overall project was completed to time and within budget. On completion it was selected from a total entry of 123 projects as the winner in the civil engineering section of the prestigious British Construction Industry Awards in October 1995.

The success of the project in meeting its overall objective of ensuring environmental enhancement and compliance with bathing water standards (Tables 2 and 3) by eliminating the 47 historic crude discharges was further recognized after its operational effectiveness had been proven, with the winning of the Norsk Hydro award for outstanding contribution to water quality improvement; a win for customers and visitors to the region and for the environment on which society depends.

Fig. 7. (top right). Penzance tunnel

Fig. 8. (middle right). Lyme Regis bathing area

Fig. 9. (right). Existing sea wall at Lyme Regis

Table 1. Penzance and St Ives project details

Element	Cost (94/5 prices)	Duration	Consultant	Contractor
Phase 1—Marazion, Mousehole, Newlyn, Penzance flow transfers, Hayle STW reconstruction, treated effluent transfer and long sea outfall at Gwithian	£68 million	March 1992 to June 1995	Pell Frischmann Water	Sir Robert McAlpine
Phase 2—St Ives pumping station and flow transfer to Hayle	£10 million	March 1994 to December 1995	Pell Frischmann Water	Sir Robert McAlpine
Resewerage contracts at St Ives and Penzance/Newlyn	£6 million	October 1992 to February 1996	Pell Frischmann Water	Sir Robert McAlpine (St Ives); Hyder Group / Trafalgar House, Trant, Clancy, TJ Brent, A Hale, E Thomas (Penzance/Newlyn)
Total cost including evaluation, design, promotion, land, legal and project management	£105 million			

Table 2. Penzance bathing water compliance

Bathing water	1988	1989	1990	1991	1992	1993	1994	1995	1996	1997	1998
Mount's Bay:											
Little Holgus	F	M	M	M	M	F	M	M	M	M	M
Heliport	F	F	F	F	M	F	F	M	M	M	M
Penzance	F	F	F	F	M	F	F	M	G	M	M
Wherrytown	F	F	F	F	F	F	F	M	M	M	M

Notes

F, failed mandatory standards; M, mandatory standards met; G, guideline bathing water quality met.

Penzance improvements were commissioned part way through the 1995 bathing season. From recent results, the Environment Agency is investigating diffuse and remote discharges from the Penzance area and its rural catchments which may from time to time affect water quality at the Mount's Bay beaches.

Table 3. St Ives bathing water compliance

	1988	1989	1990	1991	1992	1993	1994	1995	1996	1997	1998
Porthmeor	M	M	M	M	G	G	M	G	G	G	G
Porthgwidden	F	F	F	F	F	F	F	F	G	G	G
Porthminster	M	F	F	F	M	F	F	M	G	M	M
Carbis Bay—Station	M	M	M	M	M	G	M	M	G	G	G
Carbis Bay—Porth Kidney Sands	M	M	M	F	G	G	G	G	G	G	G
The Towans (Hayle)	M	M	M	M	G	G	G	G	G	G	G
The Towans (Godrevy)	N/D	M	M	M	G	M	G	M	M	G	G

Notes

F, failed mandatory standards; M, mandatory standards met; G guideline bathing water quality met; N/D, not determined.

St Ives improvements were commissioned for the 1996 bathing season. In 1998, Porthmeor beach was awarded a Blue Flag and a regional seaside award was given to Porthminster beach.

Lyme Regis

Lyme Regis is a small coastal town on the Devon/Dorset border (Fig. 8) It is set in an area of outstanding natural beauty with sites of special scientific interest both to the west and the east. The town has many listed structures and buildings, perhaps the most famous of which is the sea wall known as the Cobb which featured in the film *The French Lieutenant's Woman*. The Cobb is a grade I listed structure and its use and repair is subject to monitoring by English Heritage.

The sea has always supported the local economy and, at one time, Lyme Regis was a major trading port. The major industry now, however, is tourism and the resident population of some 5000 swells to more that 15 000 during the peak of the summer months. Lyme Regis is also well known because of the many fossils in the area, which are regularly exposed by local landslips and erosion of the mudstone and shale cliffs.

There were two major infrastructure problems. First, the town is served by a combined sewerage system which, as originally designed, discharged crude sewage direct to the sea through two short sea outfalls, one to the east and the other to the west of the town. In addition, there were thirteen combined storm overflows into the River Lim, which runs through the town to discharge into the bathing area. Consequently, there was visible pollution and debris on the beach and the bathing water failed to meet the requirements of the bathing water directive; clearly a major concern for a town where tourism is so important.

In parallel with the water company's assessment of coastal sewage treatment requirements, the local authority, West Dorset District Council (WDDC), faced the urgent challenge of stabilizing the deteriorating grade 2 listed sea walls. The sea walls were in an increasingly fragile state and following sea action during storms were often in need of emergency repairs. Additionally, the continual undermining of the sea walls was placing nearby properties at risk (Fig. 9).

The following options were evaluated

- collection and storage of stormwater flows from Lyme Regis for treatment at a foreshore works
- collection and storage of stormwater flows from Lyme Regis for partial treatment and dispersion by long sea outfall
- collection and storage of stormwater flows from Lyme Regis; pumped transfer to existing treatment works at Uplyme; works replacement; high quality effluent discharge to the River Lim
- collection and storage of stormwater flows from Lyme Regis; pumped transfer to existing treatment works at Uplyme; works replacement; high quality effluent discharge offshore by medium length sea outfall (selected option).

The reasons for the selection of options adopted for implementation were as follows

- construction of a foreshore works was impractical in the centre of Lyme Regis

Fig. 10. (above). Completed foreshore works at Lyme Regis

Fig. 11. Stormwater box culvert at Lyme Regis

Table 4. Lyme Regis project details

Element	Cost (94/5 prices)	Duration	Consultant	Contractor
Foreshore works including pumping station (Phase 1)	£8.0 million excluding West Dorset District Council funding	January 1993 to February 1995	Rendel Geotechnics	Costain Building and Civil Engineering
Resewerage—including sewer relining and pumping station and rising main from the Cobb to Gun Cliff	£2.0 million	November 1993 to May 1995	West Dorset District Council	Hannon Young
Sewage treatment works, pipelines and sea outfall (Phase 2)	£6.8 million	January 1994 to May 1995	Pell Frischmann Water	Costain Engineering and Construction (stw); TJ Brent (pipelines); Dredging and Construction (outfall)
Total cost including evaluation, design, promotion, land, legal costs and project management:	£16.8 million			

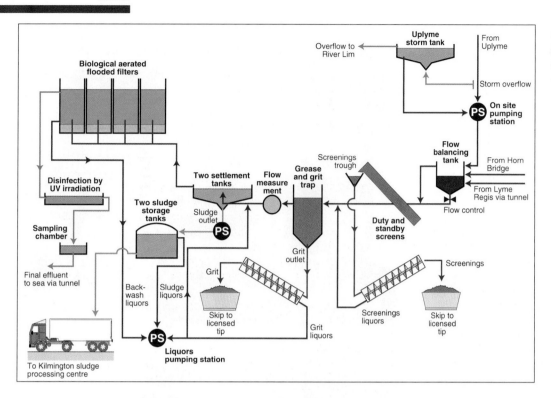

Fig. 12. Process diagram of Sleech Wood sewage treatment works

- no discharge point existed west or east of the town due to the instability of cliffs
- no inland river was available to offer adequate initial dilution for the proposed new works
- a well screened existing treatment works site at Uplyme capable of reconstruction and expansion north of Lyme Regis was available
- initial dilution and offshore dispersion criteria could be met by construction of a directionally drilled medium sea outfall from a temporary platform immediately east of the transfer pumping station and stormwater storage unit on the foreshore.

A major element of the sewage treatment proposal was the provision of a sewage collection tank at the end of the existing crude sewage discharge pipe, joint proposals were developed between the company and the district council to construct, at Gun Cliff at the easterly end of the town on reclaimed shore, a new reinforced concrete sea wall 264 m long (Fig. 10). This would have rock armour protection to the seaward face and incorporate box-section stormwater storage tanks providing a capacity for 1900 m³ of sewage storage (Fig. 11).

A new pumping station in the void between the old and the new sea wall, incorporated into the sea wall structures, enables the crude sewage collected from the tanks at the low point of the system to be pumped and transferred to a new inland treatment works through a tunnel 210 m long and 3 m dia. under the closely packed foreshore properties.

The pumping station is a wet well/dry well arrangement, pumping 65 l/s against a static head of 56 m, which required two duty pumps in series. There is 100% standby capacity. Odour control is par-

ticularly important and air from the stormwater storage tanks and pumping station is passed through a wet gas scrubbing system using hydrogen peroxide and caustic soda before venting to the atmosphere through a disguised exhaust stack—the turret in Fig. 10. The system incorporates gas detection and allows for five air changes per hour, processing vented air at a peak rate of 8·5 m³/s. The pumping station is maintained at below atmospheric pressure.

To meet local requirements the project was engineered in two phases, with the foreshore works described above being carried out in Phase 1 of the scheme (Table 4).

Phase 2 of the works consisted of transfer sewers, approximately 1·5 km long, to and from the treatment works at the nearby inland village of Uplyme, a new treatment works at Uplyme and a medium sea outfall to discharge the highly treated waste water some 600 m from the shore.

The transfer sewers comprised 300 mm dia. ductile iron pumping main and 400 mm dia. concrete gravity sewer for the return of effluent, both laid in a common trench.

The existing treatment works serving Uplyme (1000 population) was demolished and replaced with a new works of 15 000 population equivalent. The site was chosen because it was not only an existing works but it was located only 1·5 km from Lyme Regis. In addition, the site was remote from housing with extensive existing tree screening and land for expansion. The main difficulty was restricted access through Uplyme and local people were very concerned about the impact of increased vehicular movement.

The problem was overcome by buying the farm alongside the existing works, which was used main-

Table 5. Lyme Regis bathing water compliance

Bathing water	1988	1989	1990	1991	1992	1993	1994	1995	1996	1997	1998
Lyme Regis Cobb	F	F	M	M	F	F	M	M	G	F	G
Lyme Regis Church	F	F	F	F	F	N/D	N/D	M	M	M	M

Notes
F, failed mandatory standards; M, mandatory standards met; G, guideline bathing water quality met; N/D, not determined in the 1993 and 1994 bathing seasons, as the sampling point and beach access were closed for the foreshore construction works.
Lyme Regis improvements were commissioned for the 1996 bathing season. From recent results, the Environment Agency is investigating diffuse and remote discharges to the River Lim, which may from time to time affect water quality at Church Beach or at the Cobb.

ly for caravanning and camping, and providing a new access road some 500 m long from the outskirts of Lyme Regis.

The new treatment works consists of an enclosed inlet works providing flow balancing, fine screening (6 mm slots), grit removal and odour control. Primary settlement in a conventional circular scraped tank, secondary treatment by BAFF, using the Biobead process, with Minibead for solids removal, and UV disinfection.

The disinfection plant consists of 84 horizontal low pressure mercury lamps arranged in three banks of 28 lamps in a single channel with level control. All banks have UV monitors with low intensity alarms and flow pacing to vary lamp availability with variations in flow. The process is designed to achieve a 20 mg/l (BOD), 30 mg/l (SS) effluent with a 4 log reduction in coliform bacteria. The works is automatically operated, including standby power generation and SCADA links to regional headquarters.

The bulk of the work was undertaken as one contract, with UV plant as a specialist contract, direct with South West Water Services Limited. The layout is shown diagrammatically in Fig. 12.

The sea outfall was required to discharge the disinfected effluent clear of the bathing water and at a location covered at all states of the tide. The most acceptable location was at Church Cliff, but the main problem was that at low water a large area of sea-bed is exposed extending some 500 m offshore. This area is known as Broad Ledge and is of considerable environmental and scientific interest.

Excavating an open trench for the pipeline would have been environmentally damaging, weather dependent and costly. It was decided to use directional drilling to pass under the beach and Broad Ledge to emerge some 600 m from the beach in 8 m of water at low tide and well clear of the bathing areas.

A major feature of the project has been the full involvement of the public with the objective of achieving a shared understanding and ownership of the problems and solutions. The activity of the Lyme Regis Voluntary Advisory Panel (LRVAP) had been very successful in this respect, and both SWWL and WDDC wanted to build on this success.

Throughout the project an information centre was established by the company and the council overlooking the foreshore works and above the existing tourist information office. The centre contained a model of the works and an exhibition of drawings and photographs and was open 7 days a week.

As the work progressed a number of public meetings were held with client and contractors' representatives giving presentations and answering questions. There was a steady and regular release of information about progress on the work and the issue of information leaflets. The local paper, *The Lyme Regis News*, also provided regular features on the project.

Both the company and the local council believed throughout the project that the public consultation was a critical success factor in gaining local support, and this was reflected at the official opening of the foreshore works in January 1995. The opening ceremony was very well attended and some 1500 people were given guided tours through the pumping station and stormwater tanks over a period of 2 days.

The enthusiasm for the scheme demonstrated by the local community, both during construction and since the works have been completed, has been a major factor of what at the outset was seen to be a demanding and challenging civil engineering project.

The result of the combined efforts of all concerned is a legacy which future generations will undoubtedly see as an integrated feature of the historic Lyme Regis landscape, but one which successfully delivers, deep below the surface, the environmental protection so essential to today's society and to future generations (Table 5; Fig. 13).

Fig. 13. Bathers at Lyme Regis

Clearwater to Blue Flag water

D. J. Kemp, BSc, CEng, MICE

Customer opinion research by Anglian Water has shown that seawater quality at bathing beaches is a high priority. Satisfying this need, as well as meeting legal obligations under the EU bathing water directive, has involved the company in a £266 million investment programme around its coastline. This paper describes the programme, called project Clearwater, and the range of engineering solutions devised. It indicates how attention paid to marine modelling, linked to engineering work on intermittent discharges as well as the continuous discharges, has led to a successful outcome, namely the first UK water company to achieve 100% compliance at its designated bathing waters.

Proc. Instn Civ. Engrs
Civ. Engng,
Millennium Beaches
1999,
132, 28-35

Paper 11712

Written discussion closes 15 November 1999

Keywords: Sewage treatment & disposal; environment; tunnels & tunneling

Anglian Water serves the largest geographical area of the water companies in England and Wales. Extending from the Humber Estuary to the Thames Estuary, its North Sea coastline is around 4000 km in length and contains 38 EU designated bathing waters. The location of these bathing waters is shown on Fig. 1.

The challenge to Anglian Water's engineering, operational and water quality personnel was to devise and implement a capital investment programme which would deliver a phased improvement to water quality around the coastal locations leading to full compliance with the mandatory standards of the EU bathing water directive by 1998.

The investment programme, named project Clearwater, was not aimed at a set of standard emissions but rather a series of engineering solutions designed to achieve the target seawater quality at each of the unique locations. Clearwater developed into a £266 million programme which was successfully completed early and within overall budget.

Figure 2 indicates the bathing water compliance record in Anglian Water over the 10 year period 1989 to 1998. The company was the first in recent years to achieve 100% compliance in the UK. At the start of the programme the company had only 31 bathing waters, subsequent additions by government raising this to 38. Clearwater had to be flexible to respond to this.

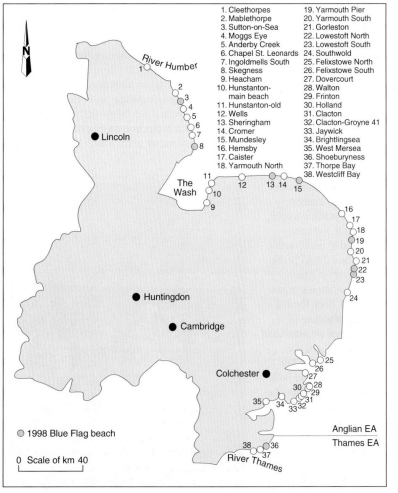

1. Cleethorpes
2. Mablethorpe
3. Sutton-on-Sea
4. Moggs Eye
5. Anderby Creek
6. Chapel St. Leonards
7. Ingoldmells South
8. Skegness
9. Heacham
10. Hunstanton-main beach
11. Hunstanton-old
12. Wells
13. Sheringham
14. Cromer
15. Mundesley
16. Hemsby
17. Caister
18. Yarmouth North
19. Yarmouth Pier
20. Yarmouth South
21. Gorleston
22. Lowestoft North
23. Lowestoft South
24. Southwold
25. Felixstowe North
26. Felixstowe South
27. Dovercourt
28. Walton
29. Frinton
30. Holland
31. Clacton
32. Clacton-Groyne 41
33. Jaywick
34. Brightlingsea
35. West Mersea
36. Shoeburyness
37. Thorpe Bay
38. Westcliff Bay

● 1998 Blue Flag beach

0 Scale of km 40

Fig. 1.Bathing waters in the Anglian Water region

Donald Kemp is production manager at Anglian Water

Encouraged by the success of project Clearwater, and recognizing the continuing needs of the coastal communities, the company in late 1997 launched a new coastal investment initiative called Blue Flag Water. This will, through investments to meet urban waste-waters treatment directive needs, together with localized projects, aim to deliver EU bathing water directive guideline standards to the region's bathing waters.

In 1998, 17 out of 38 bathing waters on the region's coastline achieved the guideline quality standard. Of these, eight had the required onshore facilities to allow the resorts to receive the Blue Flag award. The locations of the eight Blue Flag beaches are indicated on Fig. 1.

Engineering solutions

Given the freedom to create appropriate solutions, the company's engineers met their objective with a variety of cost-effective designs. These ranged from a secondary treatment works with the UK's largest ultra-violet disinfection plant at Cleethorpes (Figs 3 and 4) to a major sewerage reconstruction, preliminary treatment works and a long sea outfall at Great Yarmouth.

The full scope and diversity of the solutions is summarized in Table 1. The solutions address both the continuous discharges from sewage treatment works and the intermittent discharges from combined sewer overflows.

At many of the coastal locations, the sewerage works have not only helped to achieve required quality standards, but also greatly reduced the risk of property flooding during storm events.

Treatment works layouts have been planned with a view to future development required under the EU urban waste-waters treatment directive.

Generally, engineering designs within the Clearwater programme were developed by Anglian Water Engineering or produced by a design–construct arrangement with selected contractors. A great many civil engineering organizations were involved in the programme and Table 2 indicates principal contractors and consultants together with their main area of activity.

Coastal modelling systems

The company has developed sophisticated computer models of the coastal waters of the region. These are important tools in the management of the company's coastal and estuarine waste-water discharges and were extensively used in the design and implementation of the Clearwater programme.

Originally, the company used a set of mathematical models developed with the WRc to predict the hydrodynamics of local coastal waters and environmental impact of discharges within this area. However, as technology moved on and the company's requirements increased, a more flexible and holistic predictive tool was required to support the capital programme.

The Anglian Coastal Modelling System (ACMS)

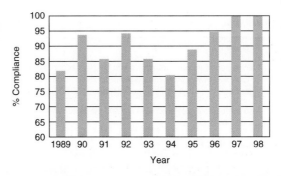

Fig. 2. Compliance of Anglian Water bathing waters against mandatory standards of EU bathing water directive

Fig. 3. (above). Aerial view of Cleethorpes sewage treatment works

Fig. 4. View of the ultra-violet disinfection unit at Cleethorpes

was developed to meet these aims. The system covers an area from Dover to Scarborough and extends around 200 km offshore. It can predict a wide range of water quality parameters and simulate the interaction and impact of discharges from sources such as rivers and industry as well as the

Table 1. Bathing water programme

Bathing water	Scheme	Project details
Cleethorpes	New secondary treatment works with UV disinfection. Four km of tunnelled interceptor/storage sewer to divert flows from existing outfall and three combined sewer outflows	Cost £50 million. Population served 100 000. Completed September 1995
Mablethorpe, Sutton-On-Sea	Improvements to sewerage system and sewage works, including UV disinfection	Cost £3 million. Population served 38 000. Completed March 1995
Moggs Eye, Anderby, Chapel St.Leonards, Ingoldmells, Skegness	Headworks improvements and provision of storm storage to reduce operation of combined sewer overflows	Cost £5 million. Population served 127 000. Completed March 1994
Heacham, Hunstanton, Hunstanton Old	New secondary treatment works with UV disinfection. Additional storm storage in sewerage system	Cost £9.5 million. Population served 25 200. Completed April 1996
Wells-Next-The-Sea	New secondary treatment works with UV disinfection.	Cost £2.0 million. Population Served 5000. Completed March 1995
Sheringham, Cromer	Diversion of flows from five short sea outfalls via 15 km of pipelines up to 600 mm diameter, to a new totally covered primary treatment works. Discharge to a tunnelled 2 km long sea outfall. 8000 m^3 of storm storage	Cost £25 million. Population served 41 000. Completed March 1995
Mundesley	Fine screening and additional storm storage	Cost £0·75 million. Population served 18 500. Completed March 1998
Hemsby, Caister, Yarmouth North, Yarmouth Pier, Yarmouth South, Gorleston	Diversion of flows from 32 estuarial outfalls, via a series of pumping stations and 20 km of pipelines, to a covered preliminary treatment works and 1·4 km long sea outfall	Cost £50 million. Population served 200 000. Completed June 1996 (project accelerated from original plan of March 1998)
Lowestoft North, Lowestoft South	Refurbishment of terminal pumping station and fine screening to all discharges	Cost £1·0 million. Population served 140 000. Completed March 1998
Southwold	Bathing water compliant. No investment required	
Felixstowe North, Felixstowe South	New secondary treatment works.Flows diverted from existing outfall via tunnelled trunk sewer	Cost £14 million. Population served 38 000. Completed June 1997
Dovercourt	New secondary treatment works and marine outfall at Harwich.New trunk/storm storage sewers to divert flows from existing two outfalls	Cost £25 million. Population served 39 000. Completed December 1997
Walton, Frinton	Additional storm storage and fine screening at five combined sewer overflows	Cost £5·2 million. Population served 31 000. Completed March 1998
Holland-On-Sea, Clacton, Clacton-Groyne 41, Jaywick	Extensive upgrading of sewerage network, additional storm storage principally in a 4·2 km, 2·1 m dia. interceptor tunnel. Fine screening at Clacton and Jaywick long sea outfalls	Cost £28 million. Population served 120 000. Completed March 1998
Brightlingsea	Additional storm storage in sewer network and fine screening on overflows	Cost £1·0 million. Population served 11 000. Completed March 1997
West Mersea	New secondary treatment works with UV disinfection	Cost £3·8 million. Population served 12 000. Completed March 1996
Shoeburyness, Thorpe Bay Southend, Westcliffe Bay	New secondary treatment works at Southend. Additional storm storage and screening to improve 28 combined sewer overflows and eliminate 22 others	Cost £41·0 million. Population served 216 000. Completed March 1998

company's own activities. The ACMS development was audited by expert modelling consultant Metoc plc.

The ACMS has been used to model the impact of the company's discharges on the region's bathing waters and to support and develop the Clearwater programme. Specifically the model has provided understanding on the influence the company's activities have on the concentrations of microbiological parameters, such as *E. coli* and *F. streptococci*, in its bathing waters.

The system is based on a triangular grid of variable mesh size (Fig. 5). It has been set up with a fine mesh close to the coast, to provide the required detail in the vicinity of the discharges and any amenity waters. Further offshore, coarser mesh sizes are used to minimize computational effort while retaining maximum accuracy in those areas where it is needed. Eight model grids are available within the ACMS, each with a different resolution and area of coverage. The coarse grid models have been developed specifically to predict the far-field and longer term (>60 days) effects of discharges using a sophisticated, depth averaged water quality model. In contrast, the fine grid models are used to predict localized effects using a dedicated midfield, particle tracking model over a period of a number of tides (typically 1–7 days)

Engineering case studies

Each of the geographical locations raised its own set of issues and problems. The following three case studies give a summary of the range of engineering challenges

- north Norfolk waste-water management scheme
- Great Yarmouth sewerage strategy
- Clacton interceptor tunnel.

North Norfolk waste-water management scheme

The towns of Cromer, Sheringham, East and West Runton and Overstrand on the north Norfolk coast were adversely affected by their short Victorian sea outfalls terminating either between high and low water or a little below low water. The two main towns of Cromer and Sheringham have designated bathing water beaches with an overall summer population of the area of 41 000.

The design solution was the pumped diversion of flows, including stored storm sewage, from five short sea outfalls to a new treatment works before discharge through a new long sea outfall. The overall cost was £25 million. The scheme layout is shown in Fig. 6.

The treatment centre was positioned to minimize pumping head while maintaining a satisfactory distance from housing and holiday facilities. The site chosen, after a lengthy planning process, was a disused gas works site. Initial works involved the removal of remaining underground plant and disposal of significant quantities of contaminated land and substances in the form of phenols and ammoniacal liquors, both being the residue from gas production.

The plant was designed for primary treatment, including coarse and fine 6 mm screens, grit removal, settlement using inclined plate separators to reduce surface area and the addition of chemical dosing at high flows and/or high suspended solids to aid flocculation. Discharge consent was set at a level of 200 mg/l for suspended solids (95% ile) with an upper limit of 600 mg/l. Due regard was given to odour control using chemical scrubbing and for noise levels by completely enclosing all operational equipment by sympathetically designed brick and tile buildings blending in with the landscape.

Figure 7 shows a schematic of the treatment plant located at Cromer. Provision was made for the secondary treatment element, which although originally not required, is now being planned for the period 2001–2003. This is due to the removal by the UK government, in 1998, of high natural dispersion areas offshore, thus requiring additional treatment in accordance with the urban waste-waters treatment directive.

The storm tanks with pumping stations were designed to contain a 1 in 5 year storm from the existing combined sewerage system, with a flow to treatment being a maximum of six times dry weather flow. These tanks needed to be positioned in key tourist areas, being the confluence of the existing sewerage network, and they ranged in size from 300 to 4000 m³.

The 4000 m³ storm tank at Cromer was con-

Table 2. Principal contractors and consultants

Contractor	Work activity
Birse Construction Ltd	Sewage treatment works at Newton Marsh, Cleethorpes, Lincolnshire
Degremont (UK) Ltd	Sewage treatment works at Wells-Next-The-Sea, Norfolk
Amec Civil engineering Ltd	Sewerage work, including large underground storm retention tanks at north Norfolk
Amec Tunnelling Ltd	Clacton interceptor tunnel, Essex
Sir Robert McAlpine Ltd	Tunnelled long sea outfall at West Runton, north Norfolk, tunnelled crossing of the River Yare, Great Yarmouth
May Gurney and Co Ltd	Sewage treatment works at Cromer, north Norfolk, sewerage work in north Norfolk, sewerage work in Great Yarmouth
Kennedy Construction Ltd	Tunnelled sewerage work in Great Yarmouth
Johnston Construction Ltd	Tunnelled sewerage work in Great Yarmouth
John Mowlem Construction Ltd	Pumping stations and sewerage work in Great Yarmouth
Galliford Ltd	Sewerage work in Great Yarmouth
A.E.Yates Ltd	Sewerage work in Great Yarmouth
Purac Ltd	Screening installations at Great Yarmouth, process plant at sewage treatment works at Harwich, Essex
Biwater Ltd	Heacham sewage treatment works, Norfolk, inlet works at treatment plant in Felixstowe, Suffolk, sewage treatment works at Southend, Essex
Jackson Civil engineering Ltd	Civil works at Felixstowe sewage treatment works
Kvaerner Construction Ltd	Tunnelled Interceptor sewer at Felixstowe, Suffolk
Clugston Civil Engineering Ltd	Civil works at sewage treatment plant, Harwich, Essex
Amey/Donelon Ltd	Sewerage works in Harwich/Dovercourt
M.J. Glancy and Sons Ltd	Sewerage works in Harwich/Dovercourt, Parkstone outfall, Harwich, Essex, sewerage works, Southend, Essex
Consultant	**Work activity**
Metoc plc	Marine modelling
Natural History Museum (marine section)	Seabed benthos and habitat surveys
Gibb Environmental Ltd	Contaminated land reclamation
Geotechnical Consulting Group	Specialist geotechnical advice
Howland Associates	Specialist geotechnical advice
Gerard Pakes Associates	Specialist tunnelling advice

structed beneath the Esplanade using diaphragm walling techniques up to 20 m in depth through gravels, clay and chalk with two rows of ground anchors to retain an adjacent cliff face. The construction is shown in Fig. 8. Electrical control equipment, odour control plant and standby generation was hidden from public view at the rear of the reconstructed Victorian shelter.

Sheringham now houses a 3000 m³ storm tank built on redundant land between the east and west promenades, previously used as a bandstand.

Construction into the sands and gravels was by a 25 m dia. caisson around 10 m in depth. The unit is now covered by an elegant brick structure housing control equipment with the remaining building handed over for local amenity use.

Smaller tanks and pumping stations were constructed at the other three resorts and were of caisson construction, two of which were finished as public amenities using a combination of local flint and brick work with rustic tiles.

The complete scheme was interconnected by 15 km of pipework ranging from 150 to 600 mm dia., with larger tank sewers up to 1·5 m dia. to attenuate flows in the towns.

The new sea outfall was tunnelled from a 29 m deep 6·9 m dia. shaft around 100 m back from the 20 m high cliff face using a 2·6 m dia. Dosco earth pressure balance tunnelling machine to a distance 2 km offshore. The level of the machine gave a minimum of 5 m cover below the sea-bed in clays and chalks, containing rounded flints well known along this coastline. The flints on one occasion eroded the cutting teeth of the machine, involving their complete replacement underground; access was gained by an adjacent tunnel constructed in heading. Within the sea outfall tunnel a 600 mm discharge pipe was placed, terminating in five diffuser risers each with four discharge ports constructed off a jack-up barge in an average 15 m depth of water.

The tunnelled solution for the outfall was chosen over other methods involving trenching of the sea-bed, partly in response to marine environment concerns expressed by local fishermen, especially those harvesting Cromer crabs. Their main worry was the amount of chalk suspension which would be generated in seawater and the harmful effect on marine life over a wide area. The tunnel option minimized this genuine concern.

This complex project built within many environmental constraints successfully produced guideline standard water quality at both Cromer and Sheringham.

Great Yarmouth sewerage strategy

The sewerage system in Great Yarmouth was designed and built in the Victorian era, and as such failed to meet modern needs. Untreated sewage was still being discharged from 32 outfalls directly into the rivers Bure and Yare until just a few years ago. These multiple discharges badly affected the quality of the towns six designated bathing waters. Figs 9 and 10 indicate the nature of the beach areas at Great Yarmouth.

A strategy was developed to intercept all of the existing outfalls and convey normal foul flows in new large-diameter sewers to a number of new pumping stations; flows are then transferred northwards to a headworks at Caister where preliminary treatment removes all objects larger than 6 mm before discharge from the 1400 m long sea outfall. Fig. 11 shows the scheme outline.

A total of 20 km of new trunk sewers and pump-

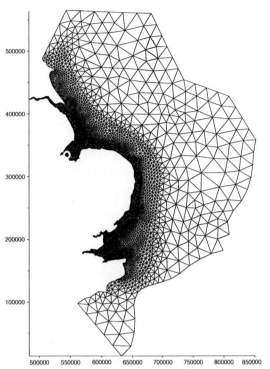

Fig. 5. Coarse grid off-shore model used to predict far-field and long-term effects of coastal discharges

Fig. 6. (below). North Norfolk scheme layout

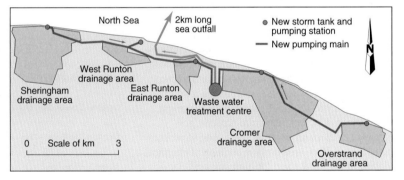

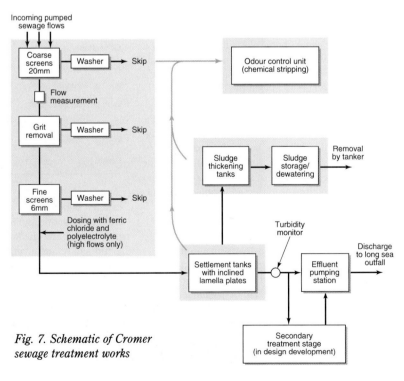

Fig. 7. Schematic of Cromer sewage treatment works

ing mains have been laid throughout the Great Yarmouth area by a mix of open cut and trenchless techniques. Much of the pipe-laying work was carried out during the winter months in order to minimize disruption to the town and its economy, which is heavily reliant on the holiday trade.

Construction works were originally programmed to continue until spring 1998. However, after discussion with Great Yarmouth Borough Council, the company agreed to accelerate the works in order to try and achieve bathing water quality compliance ahead of the 1998 target. Close co-operation with the council facilitated land purchase requirements and identified and resolved potential problems before they could affect construction progress.

Several large contracts were negotiated rather than competitively tendered, as had been planned. It was obvious that the vital 2·4 m int. dia., 420 m long tunnelled crossing under the River Yare would be the key to completing and commissioning the new network of sewers and pumping stations ahead of schedule.

The tunnel had to be driven under the ground anchors supporting the river quay walls, at a depth of 25 m, through water bearing sands and gravels.

An 800 mm dia. pumping main installed inside

Fig. 8. View of diaphragm walling construction on Cromer promenade

Fig. 9. (below left). View of Great Yarmouth beaches, looking north to pier beach

Fig. 10. (below right). View of Great Yarmouth beaches, looking south to Gorleston beach and harbour mouth

the tunnel conveys sewage flows of up to 700 l/s from the Southtown and Gorleston areas to the terminal pumping station at Bryants Quay. This pumping station also receives flows from central and south Yarmouth and pumps on over 1000 l/s to the headworks at Caister via a 6·5 km long 1000 mm dia. main. The total flow through the headworks and long sea outfall is over 1400 l/s.

The tunnel was commissioned in the summer of 1996, allowing the last of the crude sewage outfalls to River Yare to be closed. Completion of this final link in the Great Yarmouth sewerage strategy meant that all six local designated beaches achieved bathing water quality compliance for the first time. This was well ahead of the company's legal obligation of a 1998 deadline. Debris on the town beaches has been largely eliminated, a much-appreciated environmental improvement, and hundreds of properties are no longer at risk from foul water flooding. This major urban sewerage reconstruction and outfall has cost around £50 million.

Figure 12 shows the dramatic improvement in quality results on completion of the sewerage strategy. Quality levels will be further improved when a £13 million scheme for additional treatment facilities to meet the urban waste-waters treatment directive comes on stream in 1999.

Clacton interceptor tunnel

Clacton has three designated bathing waters. The main threat to quality compliance came from the intermittent discharges from around 20 storm sewage overflows issuing directly through three outfalls on to the beach area.

The project plan was to drive a 4·2 km tunnel, 2·1 m dia., parallel to the coastline, to intercept the storm sewage flows and hold them before pumped discharge to Clacton sewage works. Fig. 13 shows the scheme layout. The tunnel acts as a storm tank of 18 000 m³ capacity able to handle drainage area run off from a 1 in 5 year storm. Beyond this frequency, a screened high-level discharge weir allows a single outflow direct to the sea.

Local pressure groups, while welcoming all quality improvements, raised many issues of concern. Principally, the threat to the stability of the coastal cliffs when driving the tunnel beneath the base of the cliffs and the effect that large-scale construction would have on the economy of the town.

Additional design reviews were carried out by independent geotechnical consultants, supported by university resources using finite element analysis techniques, to verify the tunnelling proposals and satisfy local critics. At the same time engineers were able to convince residents and businesses that disruption was being designed out of the scheme.

Tunnelling finally started at the end of February 1997, some six months later than the original project plan. To recover the lost time, a partnering agreement was negotiated with the tunnelling contractor on a gain/pain share basis. Significant financial incentives were put in place to encourage early completion.

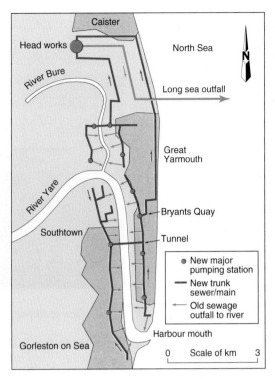

Fig. 11. Great Yarmouth sewerage layout

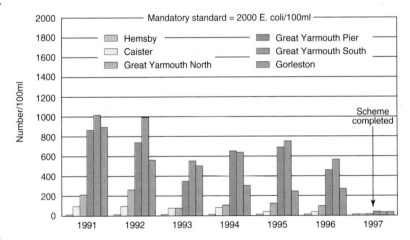

Fig. 12. Environment Agency bathing water sampling results median E.coli/100 ml 1991–1997

Tunnelling was carried out from a single drive shaft at Holland Haven, north of Clacton (Fig. 14)

A Lovat earth pressure balance machine was used to drive a 2·1 m int. dia. tunnel through London Clay at an average depth of 16 m. Prior to the summer season of 1997, 16 access shafts were constructed and reinstated to permit normal traffic movement.

Progress was impressive, particularly as anticipated water bearing gravels were not encountered. A day shift achieved 37 m in a single shift, and, on three occasions, 68 m was achieved in two consecutive shifts.

The tunnel lining was a fully gasketed bolted trapezoidal precast segment, constructed to ensure water tightness to prevent lubrication of the layers of clay at the base of the cliffs.

An extensive array of inclinometers was installed between the cliff edge and tunnel. Monitoring revealed no movement. Likewise, all nearby properties were surveyed before and after the project, with no observable damage.

A 400 m long tunnelled connection using 2·8 m dia. bolted segments was hand-driven from the main drive shaft to the sewage treatment works.

Substantial completion of tunnelling was achieved 12 weeks early in December 1997, with an excellent safety record, at a cost of £11 million. Ancillary works were completed by March 1998 in readiness for the 1998 bathing season.

Seawater quality at Clacton was measurably improved and residents observed a significant reduction in sewage-related litter along the shoreline, a welcome aesthetic and environmental benefit.

The sceptical residents admitted that they had witnessed a well-engineered project.

Public relations

A perceived failing of the engineering profession has been its reluctance to advertise its work. It was to address this potential weakness that the concept of project Clearwater was devised, an umbrella title covering all the bathing water investment programme.

Clearwater was designed as a communications programme to spell out, in simple language, the environmental and service benefits of this £266 million investment. An associated logo, suggesting sunny days by the sea (Fig. 15), was used extensively in communications material.

The messages were delivered through open days, roadshows, newsletters, advertisements and media releases about the priorities and progress. Local issues were highlighted in the context of the regional programme. Residents were informed about how the engineering activities would impact on them.

Local schools were involved, with competitions such as 'name the tunnelling machine'. Very often this was a first and formative contact with the world of civil engineering.

Compliance manuals

Maintaining high quality levels in seawater on completion of engineering projects is no easy matter. Removal of major sources of bacteria sometimes only serves to reveal a previously unknown, but now significant, input which can compromise quality.

To aid operational personnel in monitoring the situation, Anglian Water has developed a series of 17 compliance manuals covering, in geographical groups, details of all the continuous and potential storm discharges which may affect designated bathing waters. These discharges have been equipped with telemetry monitoring to develop patterns of response in rain events.

This not only aids understanding of the new infrastructure and raises operational awareness, but also gathers information to defend the company in the event of a failing sample.

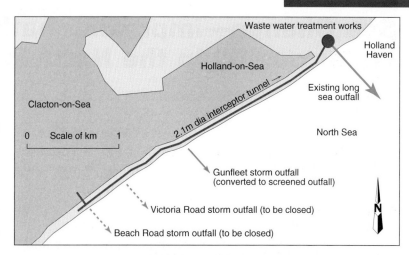

In the near future these manuals will assist the investigations for the next tranche of investments which will be required to deliver the Blue Flag water policy, namely EU bathing water guideline standard at all of East Anglia's bathing waters.

Acknowledgements

Grateful acknowledgement is given to the following Anglian Water staff for their assistance with this paper: Paul Hickey, Monica Greenwood, Kevin Whybrow, Owen Leonard, Paul Mitchell, Adrian Henderson.

Fig. 13. (top). Clacton interceptor tunnel

Fig. 14. (above). Aerial view of the drive shaft site at Holland Haven, Clacton

Fig. 15. The PR logo

Sea Change—improving bathing water quality in the North West

S. Rasaratnam, BSc(Eng), MSc, MBA, MIWEM

*Proc. Instn Civ. Engrs
Civ. Engng,
Millennium Beaches*
1999,
132, 36-44

Paper 11713

*Written discussion closes
15 November 1999*

*Keywords: Sewage
treatment & disposal;
environment; tunnels
& tunneling*

In response to the EU bathing water directive, North West Water mounted a £500 million-plus initiative along its considerable coastline to help bring bathing waters—including Blackpool, the UK's most popular resort—up to European standards. The operation, known as Sea Change, comprised many and varied waste-water projects, including in the Fylde the world's largest submerged biological contactor treatment plant, and the longest MDFE outfall pipe. This paper describes the challenges faced and the solutions adopted in completing this major engineering programme in time for the 1996 bathing season, with particular reference to the Fylde scheme and the ongoing programme of further investigation and investment.

The north west coast of England is renowned for its seaside resorts, established in the Victorian era to serve the Lancashire textile mill towns. Blackpool is world-famous as a fun family resort (Fig. 1) and attracts 17 million visitors a year, more than any other location in the UK or even the whole of Greece. Southport (Fig. 2) and Lytham St Annes are well known more genteel resorts. Morecambe, close to the Lake District, is famous for its shrimps and its sunsets across the bay. To the north, along the Cumbria coast (Fig. 3) in the Lake District, there are a number of small coastal towns which attract visitors during the summer.

Sadly, the legacy left behind by the Victorians is more than just famous resorts. The sewerage systems they installed discharged untreated waste-water directly offshore, and as populations have grown these systems have no longer been able to cope. Sewage flooding and discharges to the shore during storms have all been a common occurrence.

This has now changed dramatically through the completion of North West Water's £500 million coastal clean-up programme. Operation Sea Change—as the programme was called—covering over 600 km of the north-west coast from the Scottish border to the Mersey estuary, is one of the biggest environmental programmes of its type in the UK, and was engineered by Bechtel Water Technology.

The programme included major schemes at Southport, Lytham St Annes/Preston, the Fylde

Fig. 1. Blackpool beach on the Fylde coast—a resort more popular than all of Greece

Shanthi Rasaratnam is special projects manager at North West Water

coast (Blackpool and Fleetwood), Morecambe,
Barrow, Workington/Maryport and a number of
smaller projects along the Cumbria coast in the
Lake District. A total of over 60 distinct engineering
projects were developed involving 17 km of waste-
water tunnels, 14 new or upgraded treatment
works, more than 35 new pumping stations and
eight new sea outfalls (Fig. 4).

History

A total of 33 bathing water locations were designated in 1987 for the first time in the North West to
comply with the EU bathing water directive of 1976
(Fig. 5). The delay was due to a change of interpretation of the directive that initially did not result in
any designations.

The designated waters of Blackpool and
Southport were required by the government to be
tackled first as a high priority, to reflect the high
populations served and their importance for
tourism. The Fylde coastal waters improvement
scheme serving Blackpool and the Southport
scheme constitute an investment of almost half the
cost of the total programme.

In March 1991, ongoing plans were overtaken by
the announcement by the government's environment secretary that the schemes being considered
for bathing water compliance should incorporate
the EU urban waste-waters treatment directive due
to come into force in the year 2000. A revised programme of work was developed by the company, in
consultation with the Department of the Environment, incorporating the urban waste-waters directive. The EU accepted a time dispensation to
recognize the increased complexity of the schemes
incorporating higher levels of treatment. All
schemes had to be substantially complete by the
1996 bathing season except Morecambe, which
was to be completed by the 1997 bathing season.

Fylde scheme

Planning

Prior to the March 1991 announcement by the
environment secretary, the industry adopted the
concept of marine treatment—that is, allowing part
of the treatment stage to be effected in the marine
environment through breakdown of bacteria by

Fig. 2. Southport beach

Fig. 4 North West Water region showing locations of coastal treatment plant and outfalls.

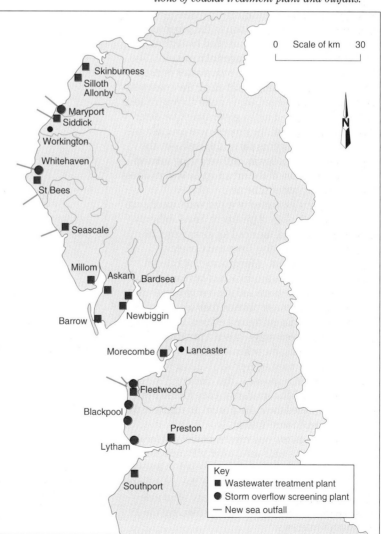

Fig. 3. The Cumbria coast in the Lake District

sunlight, salt water and effective dispersion and dilution. This was the basis of the initial feasibility studies on the Fylde.

Based on the marine treatment principle, investigative work commenced in 1988 on the Fylde coast, including hydrodynamic modelling with verifications using ocean surface current radar and floats, dispersion modelling with verification using tracer bacteria, bottom sediment sampling, baseline ecological surveys and WASSP modelling of the onshore sewer network.

The models developed were integrated and used for simulation of the system to predict the outcome of various proposals considered. The techniques developed for the Fylde coastal waters scheme (with the assistance of the Water Research Centre) and the analysis of options pioneered the methodology that subsequently became widely used throughout the UK for similar projects.

Extensive consultations took place on the proposals being considered, including public exhibitions and meetings with pressure groups and local authorities. The modelling work indicated that long outfalls discharging to suitable locations after partial treatment would be sufficient to enable the designated bathing waters to meet the EU mandatory standards.

There was public opposition to the proposals, led by pressure groups that demanded biological secondary and even tertiary stages of treatment. The planning proposals based on stormwater management using storage tanks built along the coast with a discharge of screened waste water through a long outfall were referred to the government for determination in 1990. But the process was overtaken—and the public concerns partly diffused—by the new government policy in 1991 requiring immediate implementation of the urban waste-waters directive on bathing water schemes.

The company examined new proposals incorporating secondary biological treatment. Local authority planning permission, the need for environmental impact studies under the Town and Country Planning (Assessment of Environmental Effects) Regulations 1988, the acquisition of land and the objections from pressure and interest groups—either not wanting a treatment plant on their 'doorstep' or wanting more stringent treated effluent discharge standards than necessary—posed a particular challenge to completing the scheme in time. This project was perhaps the most high-profile bathing water project in the UK. In response to this challenge, the company set up a public consultative body for the Fylde coastal area, called the Fylde Forum, to reinforce the concept of public participation.

The unique forum had representatives of local authorities over a wide area, local MPs and MEPs, and other interest groups. The company's director of planning chaired the forum, with the acceptance of the other members, in the capacity of an independent chairman rather than representing the company's interests.

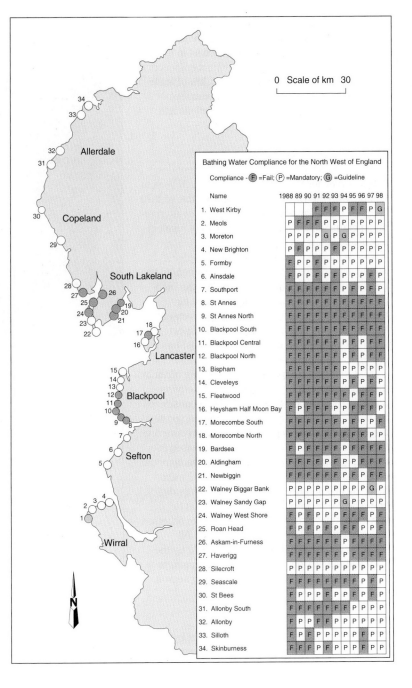

Fig. 5. Map of designated bathing water locations showing compliance with EU bathing water directive

Numerous proposals with associated costs were tabled at regular meetings; these were eventually whittled down to just three. The final decision on the choice of scheme was left to the company because the proposals crossed local authority boundaries and local councillors and interest groups did not want to be associated with a solution that might have had political implications. The selected scheme is as shown in Fig. 6.

Public opposition then transferred to the local population/council in which the newly proposed treatment works was to be located, namely Wyre Borough Council. The council's planning requirement was that the treatment works serving a population equivalent of 330 000 be completely covered to resemble visually a high-class industrial park and to contain potential odour problems within the works. This was to avoid the image of Fleetwood being tarnished by having a large waste-water treatment works serving the whole Fylde coast next to the main access road to the town. Despite this approach being adopted, after much debate, presentations and public exhibitions, the planning authority referred the proposals to the government, which called a public inquiry. The inquiry found in favour of the company in February 1993.

Design

Good design minimizes construction impacts, particularly on major schemes in large coastal conurbations such as Southport, the Fylde coast and Morecambe. The particular problems of disruption caused and the planning restraints imposed were recognized on the Fylde.

An elegant solution was developed to deal with stormwater management to minimize the frequency of stormwater discharges. Initially, significant storage had to be provided—in the order of 70 Ml in total—at three locations adjacent to existing coastal pumping stations. This would have created major disruption in constructing large storage tanks (one the size of a football pitch), pumping stations and surface pipelines along the entire length of the built-up coastal strip of Blackpool, which in summer has the second densest population in the UK. Instead, the designers proposed a 2·8m dia., 11·8 km tunnel solution, which provided both the storage and the gravity transfer of waste water to the treatment plant at Fleetwood. This approach minimized both disruption of the local community during construction and environmental intrusion of the finished works.

The need to cover the treatment works completely necessitated a compact footprint for the plant, and an imaginative architectural style of buildings covering the plant to avoid excessive costs (Fig. 7). A complex ventilation system and a chemical scrubbing plant was designed to maintain a safe atmosphere within the buildings and to clean the air

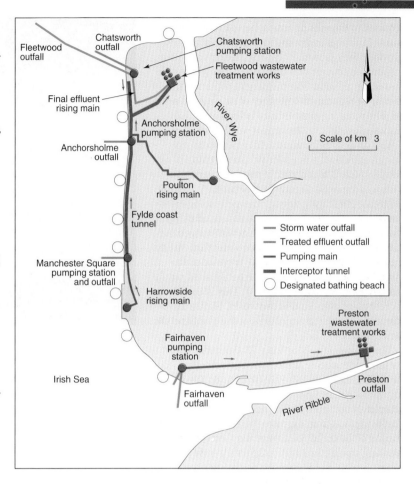

Fig. 6. Plan of the Fylde coast scheme

Fig. 7. Fylde covered treatment plant

released to the outside atmosphere. Process plant within buildings and prone to release odours was provided with secondary covers. The air contained within these covers was treated separately from the rest of the air within the buildings to optimise the design by reducing the number of air changes needed. The resulting plant (Fig. 8) is the largest fully enclosed waste-water treatment works in the UK.

The process selected to achieve the compact design avoided the need for primary treatment by adopting fine 3 mm screening and grit removal. This was followed directly by a secondary biological treatment stage using submerged biological contactors (Fig. 9). The process consisted of 5·5 m dia. plastic circular rotating packs with a large surface area, immersed in the waste-water flow stream into which process and drive air is provided to allow the micro organisms to be established on the interstices of the packs (Fig. 10). This is the largest plant of its kind in the world. Figure 11 shows an annotated aerial view of the plant and the processes.

The design of the 5 km long, 1300 mm dia. outfall for the Fylde scheme at Fleetwood presented particular difficulties due to there not being a suitable stringing site for a pulled outfall, the coastline being in a completely built-up area. The adopted solution was to float in to place, from a remote storage location, 500 m lengths of the outfall, joint them and then sink the resulting pipeline into a pre-excavated trench.

Construction

David Bellamy, the well-known environmentalist and broadcaster, publicly launched the £500 million Sea Change programme in Blackpool, in May 1994 (Fig. 12). This was just before major construction work commenced on the Fylde.

The construction strategy adopted was to divide

Fig. 8. Final settlement tanks inside the Fylde covered treatment plant

Fig. 9. Installation of submerged biological contactors at Fylde

the scheme into specialist contracts for the outfalls
and major tunnelling work and to award the contract for the state-of-the-art fully covered treatment
plant to a design/management contractor. The
management contractor would co-ordinate the work
of about 16 main contractors for various elements
of civil, mechanical, electrical, building and ICA
work. This approach was adopted because of the
tight deadline (and hence the resource level needed) for completing the scheme created by the delay
from the public inquiry.

At some of the larger coastal towns of Blackpool,
Southport and Morecambe where major new tunnel sewers, pumping stations and outfalls were
being constructed as part of the overall scheme, it
was inevitable that inconvenience and some nuisance to local residents would be caused. From the
experience of opposition to some of these schemes
at the planning stage, the importance of keeping
local residents informed of construction implications and its progress was realized. The response
was to provide (normally before the commence-
ment of construction) information centres at strategic locations manned by dedicated staff. At these
centres the proposals were displayed and illustrated
leaflets made available. This approach paid-off by
reducing the antagonism to the construction work,
and led residents to be more accommodating in the
hours of work allowed to the contractors. It was
important that the company staffed these centres in
order to gain credibility from the public.

The construction programme was challenging by
any yardstick. Value of work in the order of £150
million on the scheme had to be completed over a
construction period of 18 months. At its peak the
weekly value of work carried out was over £2 million. At the peak of construction there were over
500 workers on the treatment works site. The co-
ordination of the interface between contractors,
dealing with the knock-on effects of delays of one
contract on another and the re-programming and
selective incentives offered to contractors to make
up lost time on critical elements, was a constant
challenge to the whole project team

The implementation of the outfall project was particularly novel. The MDPE pipes of 1300 mm dia.
were manufactured in Norway and towed across the
North Sea to the Fylde coast in bundles of five pipes
each 500 m in length (Fig. 13). On one occasion one
of the pipes became detached while being towed-in
due to stormy weather and ended up on the shore
badly damaged. The pipes were moored within the
Wyre Estuary away from the impact of stormy
weather until the time to be towed out, connected up
end to end on a barge at sea (Fig. 14) and then sunk
into position in the pre-excavated trench by weight-
ing the pipes with concrete collars. The jointing
using a unique Aquagrip type coupling was specifically developed for this outfall (Fig. 15). The 5·25 km
outfall is the longest MDPE outfall in the world.

However, the greatest challenge to meet the
deadline was in the tunnelling work. Driving of the

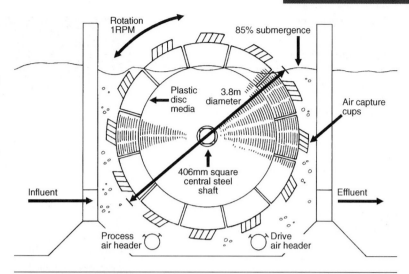

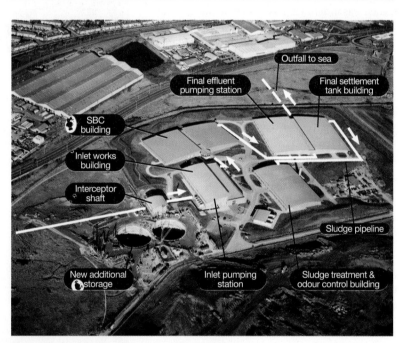

*Fig. 10. (top).
Operation of sub-
merged biological con-
tactors*

*Fig. 11. (above).
Aerial view and
process flow at Fylde*

*Fig. 12. David
Bellamy at the Sea
Change launch*

41

main transfer/storage tunnel, 11·5 km long, commenced using two earth pressure balance tunnelling machines, one being launched from the treatment works and the other starting half way along the length of its route, but both moving in a southerly direction (Fig. 16). Initial slow progress was experienced due to teething problems with the tunnelling machines, and the difficulty of providing a machine that could cope with varying ground conditions—a combination of boulder clays and running sand. The machines broke down and the cutting heads had to be replaced frequently.

After six months only 5% of the total tunnel had been completed. The problems persisted and the completion date became critical. The construction strategy including contract incentives was reviewed with the contractor and a third tunnelling machine incorporating improvements from the experience gained was ordered in haste from the manufacturers in Canada, so that three tunnelling machines could be in operation. Spare cutting heads were also ordered to reduce maintenance time. As time was of the essence, the machine weighing 100 tonnes was assembled at the works and air-freighted to the UK. The only airport capable of receiving the loaded transporter was Stansted, 320 km from Blackpool. This may have been the first time that a tunnelling machine of this size has been assembled and transported by air to meet a time deadline.

Staffing on site was strengthened and daily monitoring of progress was undertaken. Key decisions, such as replacing cutting heads and servicing the tunnelling machine at the next shaft rather than risk having a breakdown halfway between shafts, had to be constantly taken to make up lost time. The completion date, which was about 3 months over the planned deadline, was nevertheless a tremendous engineering achievement given the initial slow progress, secured by co-operation between all parties concerned.

Commissioning

The numerous state-of-the-art treatment plants and associated stormwater management network systems being commissioned along the coastline over a relatively short period was a new experience for the company. This was due to the significant upgrading required to waste-water systems along the coast; a waste-water system previously associated with minimum levels of treatment and storm management.

Challenges to commissioning the schemes in time were inevitable given the latest technologies being adopted for processes and the limited experience of operating storm management systems with significant storage just upstream of the treatment works.

The difficulties were most pronounced in the

Fig. 13. (above right). Outfall pipes floated in 500 m lengths from Norway

Fig. 14. Outfall pipes being installed

treatment plant serving the Fylde area at
Fleetwood. Within a short time of turning flows
into the new plant, the fine-screening plant became
clogged with slugs of screening debris and sludge;
this was most pronounced after intense storms.
The knock-on effects on sophisticated ICA systems
used to protect the plant caused the inlet works to
flood within a very short time.

The phenomenon causing the sudden clogging of
screens—although not fully understood—has been
attributed to the flat nature of the sewer network
and the large installed tunnel storage capacity.
Anticipation of storms and controlled pumping of
flows into the works have now resolved the problem. In addition, the sludge produced from the
highly loaded SBC process was unstable, leading to
the release of hydrogen sulphide which caused
odour problems at Preston, where the sludge is dewatered ready for landfill.

Whereas extensive preparation and planning had
taken place for the critical commissioning phase,
the scale of difficulties was not anticipated.
Commissioning responsibility was given over to the
consultancy. It soon became apparent that the level
of staffing for commissioning and operation needed
strengthening, with less reliance on automated systems. In addition, the roles and responsibilities of
the water company, the consultancy and the management contractor needed further clarity. A dedicated company commissioning manager was
appointed to lead its interests, and representatives
from each organization were given more clearly
defined roles and responsibilities. This approach
was extremely effective in resolving problems and
was soon implemented on other projects approaching the commissioning phase.

The commissioning team, with extensive data

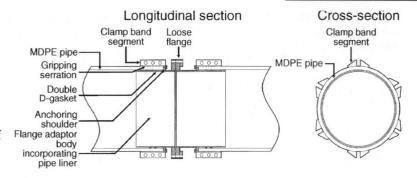

Longitudinal section

Cross-section

Fig. 15. Cross-section
of outfall pipe joint

gathering and modelling of the process, investigated the problem of unstable sludge due to the high
loading on the submerged biological contactor
process. The problem has been resolved by means
of step-feeding of the contactors to even the loading
rate, and the provision of additional process air and
chemical conditioning of the sludge. The process
understanding and insights developed as a result
have been considerable.

Performance

Since the summer of 1996, following substantial
completion of the Sea Change programme, the new
assets created have operated successfully, meeting
and in some cases significantly exceeding the discharge standards set by the Environment Agency.
Despite this performance, of the 33 bathing waters
originally designated in the North West in 1996,
1997 and 1998, 13, 17 and 13 locations, respectively,
failed to meet the mandatory EU standard in the
bathing season (Fig. 4). There has been an
improvement in bathing water quality at these locations, but not quite enough to meet the stringent
standards consistently.

Fig. 16. Earth pres-
sure balance tun-
nelling machine at
start of transfer/stor-
age tunnel drive

The reasons for less-than-total success are being continuously investigated. Extensive investigative work on the Fylde coast was carried out by the Environment Agency supported by the company and the University of Leeds. The studies looked at the statistical links between environmental factors and bathing water quality and the bacterial inputs to the Ribble estuary from both riverine sources and point sources (waste-water treatment works, stormwater overflows and pumping station overflows).

Further dye and bacterial tracer studies were carried out to prove connectivity further, in particular from inland sources. CCTV surveys of the coastal sewers and Ariel infrared surveys of the coastline were carried out to look for unidentified outfalls—but none were found. What this problem highlights is the complex and diffused nature of pollution in the local marine environment due to estuarial impacts and other associated inland sources (e.g. agricultural practices and possible illegal discharges).

The problem areas, not surprisingly, are concentrated in the eight designated locations of the Fylde coast, which straddles two estuaries and is the location of a significant part of the total population load of the Sea Change programme. There are also random failures along the Morecambe Bay and Cumbria coastlines from year to year. Inevitably, European and government interest has been focused on the Fylde coast which, in Blackpool, contains the best-known seaside resort in the UK.

The pressure to improve performance has led the Environment Agency, supported by the Department of the Environment, Transport and the Regions, to ask the company to implement the following precautionary measures.

- Reduce the frequency of storm discharges along the Fylde coast to three per bathing season, from the current nine to twelve discharges on average, by providing further storage of 90 Ml in the Blackpool sewerage network and 67 Ml at the Preston treatment works.
- Provide UV treatment to other discharges possibly influencing the Fylde coastal waters from the Ribble estuary, such as the treated effluent from the Preston treatment works, the treated effluent from the Southport treatment works and some other smaller estuarial works.

The proposals are planned for completion by the 1999 bathing season. The single UV contract will be the largest in the world. These measures should further reduce the bacterial inputs from the estuary suspected of having an impact on the bathing waters, particularly in the southern half of the Fylde coastal waters at Lytham St Annes and Blackpool.

Further studies are being undertaken by the Environment Agency with the assistance of the company to understand the cause of failures. These include the development of a more sophisticated hydrodynamic model for the Ribble estuary, tracer studies using novel DNA labelling techniques and bacterial 'finger printing' to try to establish the source of the bathing water pollution.

Summary

The planning stage of the Fylde project - the largest and most complex scheme undertaken in the Sea Change programme - was particularly difficult due to local community opposition to the location of the waste-water treatment plant and the demands for higher standards of treatment. This was eventually resolved through pubic inquiry. The sensitive treatment of the flora and fauna in some areas also emphasized the increasing importance attached to environmental matters.

On the Fylde scheme, innovative designs in the form of tunnelled pipelines for transmission and storage, and the novel techniques to float in outfalls to avoid launching sites in restricted areas, stand out. The use of intensive non-traditional treatment processes, such as submerged biological contactors, reduced footprints to allow economic covering of process plant units, demonstrate the influence of planning and environmental concerns in the design process. The outcome in the Fylde was the creation of the largest fully covered plant in the UK.

The biggest challenge during construction was the limited time available, partly due to planning delays and prescriptive completion dates agreed with the EU. The race against time on the tunnelling work on the Fylde, and the response of all parties involved, was a key example of how teamwork and partnership can lead to notable achievements. The transport of outfall pipes from Norway to the Fylde coast, by floating them across the ocean, was a unique and impressive approach.

Commissioning of large integrated schemes with relatively new innovative treatment processes created its own share of problems. The response to this difficulty was the creation of dedicated commissioning teams with representatives from all parties to manage and implement the process effectively. The creation of unstable sludges from the treatment process and the control of complex storm management systems and its effect on the treatment process were particular challenges during the commissioning phase.

Despite significant improvement, the inability to comply fully with EU bathing water directive standards has revealed the complex nature of pollution in the marine environment. The Environment Agency is further investigating the reason for this phenomenon, which is not clearly understood. In the meantime, further precautionary measures are being implemented by the company on the request of the DETR and the Environment Agency, for completion in the summer of 1999.

Finally, the completion of the £500 million Sea Change programme in time and within budget, while circumventing the numerous obstacles along the way, is a tremendous engineering achievement that all participants on the programme can look back on with satisfaction.

Northumbrian Water's investment to end sea-disposal of sewage sludge

S. Coverdale, BTech, CEng, MICE and A. N. Hill, BSc, MIMechE

Proc. Instn Civ. Engrs
Civ. Engng,
Millennium Beaches
1999,
132, 45-52

Paper 11714

*Written discussion closes
15 November 1999*

*Keywords: Sewage
treatment & disposal;
environment; waste
management & dispos-
al*

Under EU legislation for bathing waters, Northumbrian Water has invested
£150 million to provide environmental improvements for the 34 designated
bathing beaches along its 180 km coastline. Now, to comply with urban
waste-water rules, the company is spending another £200 million to provide
further treatment to coastal discharges. All this will lead to a 200% increase in
sewage sludge over the next six years, none of which can now be dumped at
sea. The company has thus built a state-of-the-art £70 million regional sludge
treatment centre on Teesside which turns the sludge into dried fertilizer pellets. In the near future it is Northumbrian Water's intention to use the dried
sludge to generate fuel gas, making the centre virtually self-sufficient.

As the appointed sewerage undertaker for north
east England under the Water Industry Act 1991,
Northumbrian Water has a statutory obligation to
provide a system of public sewers and deal effectively with their contents. The 2·6 million population
served by the company lives predominately along
or near the North Sea coastline, in particular, within
the three major conurbations of Teesside, Tyneside
and Wearside.

Not surprisingly, the sewerage network inherited
by the company relied heavily on sea disposal of
sewage—marine treatment—and the company's
capital investment programme has been and is
dominated by massive engineering works to clean
up these discharges.

At a cost of £150 million, each of the 34 bathing
waters, from Berwick in the north to Saltburn
(Fig. 1) in the south, have seen a combination of
storm attenuation, flow interception/transfer, headworks and long sea outfalls to allow each bathing
water to comply with EU legislation.

Figure 2 shows the location of each of the
region's bathing waters and their
respective performances.
Although there has been a consistent improvement over the last
nine years, in 1998 eight of the
bathing waters in the region
failed the water quality tests carried out by the Environment
Agency, compared to two in 1997.

Samples are taken during the
summer regardless of the weather, whereas many EU countries
avoid taking samples when it is
very wet. The summer of 1998

was exceptionally wet both in terms of severity and
frequency of rainfall, which increased the impact of
diffuse pollution and discharges from storm sewerage overflows into watercourses. Furthermore,
some of the adverse results are the consequence of
events outside the company's control, such as the
River Aln changing its course and a large discharge
of pig slurry into the watercourse crossing the
beach at Saltburn.

The government is currently considering setting
new targets for bathing water quality; this will
require a lot more investment, particularly in
improving emergency overflows. The company will
make whatever investment the government decides
is necessary.

A list of the principal consultants and contractors
involved in the bathing waters programme is set
out in Table 1.

Costing £200 million, a second tranche of works,
instigated by the implementation of the EU urban
waste-water treatment directive, will provide eight
new sewage treatment works along the coast at

*Steve Coverdale is pro-
gramme manager at
Northumbrian Water
Limited*

*Fig. 1. Saltburn beach—south-
ernmost beach in the region*

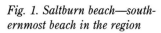

*Tony Hill is project
design manager at
Entec*

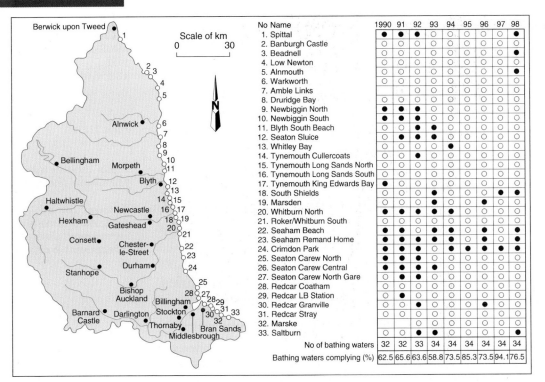

No	Name	1990	91	92	93	94	95	96	97	98
1.	Spittal	●	○	●	○	○	○	○	○	●
2.	Banburgh Castle	○	○	○	○	○	○	○	○	○
3.	Beadnell	○	○	○	○	○	○	○	○	●
4.	Low Newton	○	○	○	○	○	○	○	○	○
5.	Alnmouth	○	○	○	○	○	○	○	○	●
6.	Warkworth	○	○	○	○	○	○	○	○	○
7.	Amble Links				○	○	○	○	○	○
8.	Druridge Bay	○	○	○	○	○	○	○	○	○
9.	Newbiggin North	●	●	●	○	○	○	○	○	○
10.	Newbiggin South	●	●	●	○	○	○	○	○	○
11.	Blyth South Beach	○	○	●	●	○	○	○	○	○
12.	Seaton Sluice	○	●	●	●	○	○	○	○	○
13.	Whitley Bay	○	○	○	○	●	○	○	○	○
14.	Tynemouth Cullercoats	○	○	●	○	○	○	○	○	○
15.	Tynemouth Long Sands North	○	○	○	○	○	○	○	○	○
16.	Tynemouth Long Sands South	○	○	○	○	○	○	○	○	○
17.	Tynemouth King Edwards Bay	●	○	○	○	○	○	○	○	○
18.	South Shields	○	○	○	●	○	○	○	●	●
19.	Marsden	○	○	○	●	○	○	●	○	○
20.	Whitburn North	●	●	●	○	●	○	○	○	○
21.	Roker/Whitburn South	○	○	○	○	○	○	○	○	○
22.	Seaham Beach	●	●	○	●	●	○	●	○	●
23.	Seaham Remand Home	●	●	○	●	●	○	●	○	●
24.	Crimdon Park	●	●	●	○	●	●	●	○	●
25.	Seaton Carew North	●	●	●	○	○	○	○	○	○
26.	Seaton Carew Central	●	●	●	○	○	○	○	○	○
27.	Seaton Carew North Gare	○	●	●	○	○	○	○	○	○
28.	Redcar Coatham	○	○	○	○	○	○	○	○	○
29.	Redcar LB Station	○	●	○	○	○	○	○	○	○
30.	Redcar Granville	○	○	●	○	○	○	●	○	○
31.	Redcar Stray	○	○	○	○	○	○	○	○	○
32.	Marske				○	○	○	○	○	○
33.	Saltburn	○	○	●	●	○	○	○	○	●
	No of bathing waters	32	32	33	34	34	34	34	34	34
	Bathing waters complying (%)	62.5	65.6	63.6	58.8	73.5	85.3	73.5	94.1	76.5

Fig. 2. Northumbrian Water region showing designated bathing waters and compliance with mandatory standards of EU bathing water directive

Amble, Newbiggin, Cambois, Hendon, Seaham, Horden, Seaton Carew (Fig. 3) and Marske by the Sea. Principal consultants and contractors delivering this ongoing programme are listed in Table 2.

The environmental improvements are achieved by using sewage treatment works that will produce sewage sludge. Currently, the region produces approximately 660 000 m³ (equivalent to 33 000 t dry solids) of sewage sludge per year, 60% being disposed to sea, 30% to landfill and 10% to agriculture.

The disposal of sewage sludge is closely regulated and is carried out in accordance with UK and EU legislation. New EU legislation (principally the urban waste-waters directive) will cause the amount of sludge from the region to more than double over the next six years (Fig. 4). At the same time, the directive bans the disposal of sludge to the sea—the company's main sludge disposal route—from the end of 1998, leading to a third tranche of work for the coastal clean-up.

The company has considered many options in the preparation of a sludge strategy to meet this latter challenge. The prerequisites of the strategy were that it had to be safe, secure, flexible, economic, environmentally acceptable and be able to serve the needs of both the company and the region for the next 20 years. The strategy was developed by

Table 1. Bathing water programme—principal consultants and contractors

Consultants	Contractors
Entec	Balfour Beatty
Montgomery Watson	Birse
	Byzak (Kennedy)
	Harbour and General
	Johnstons
	Mowlems
	Sir Robert McAlpine
	SPP
	Van Ord ACZ
	Weir Pumps

Table 2. Coastal programme— principal consultants and contractors

Consultants	Contractors
Babtie	Byzak
Binnie Ferguson	Balfour Beatty
McIlveen	Degremont
Cundall Johnston	Mowlems
Entec	
Montgomery Watson	

Table 3. RSTC phase 1 contract awards

Contract	Contractor
Combined heat and power plant	Tuma Turbomach (commissioning)
Dredging (jetty)	Dredging International
Drying plant	Andritz AG
Drying plant building	J Mowlem
Electricity (HV supply)	Northern Utilities Services Ltd.
Enabling works	Harbour and General
Gas supply	Enron
Jetty construction	Nuttall
LV electrical and instrumentation	N G Bailey
Main civils	Clugston Construction
Mechanical services	Siemens
Piling	Bachy
Principal contractor (CDM)	Sir Robert McAlpine
Site security	Barrier
Software/SCADA	Blackburn Starling

the company using specialist consultants including ERM, Binnie and Partners and Entec.

Sewage sludge utilization strategy

After careful consideration and development of numerous options over a period of two years, the company adopted the following sewage sludge utilization strategy as the strategic best practicable environmental option for the treatment and disposal of the majority of the future sludge arisings in the region.

- Sludges will be brought together, at Bran Sands on Teesside, for dewatering and drying.
- Thermal drying is the chosen sludge processing technique, and this has a proven track record.
- The ultimate design capacity of the plant will be 90 000 t dry solids (tds) per annum. This comprises 70 000 tds of sludge from domestic and trade waste from treatment works around the Northumbrian region and 20 000 tds of sludge from domestic and trade waste from the Tees estuary environmental scheme effluent treatment works at Bran Sands. This latter capacity will support future industrial development on Teesside through the provision of waste treatment facilities. Waste liquors from the regional sludge treatment centre (Fig. 5) will be treated at the adjacent effluent treatment works.
- The total design capacity of the sludge treatment centre will be developed in several phases. The first phase was commissioned in 1998 at a cost of £70 million to provide a 50 000 tds a year treatment capacity. The second phase, programmed to be commissioned in 2000, will increase plant capacity to 75 000 tds a year. Subsequent phases will provide additional capacity to cope with future sludge increases.
- It is intended that the plant will produce a range of dried sludge products for use as a fuel, in a variety of soil conditioner markets (e.g. agriculture, forestry and land reclamation) and as a carbon replacement (e.g. in the manufacture of bricks or steel). As the material produced may be used as a fuel, the plant will be licensed under integrated pollution control legislation issued and monitored by Environment Agency. If beneficial outlets cannot be utilized, the dried sludge will be landfilled (taking up a reduced volume).
- The majority of sludges would be shipped to Bran Sands from sewage treatment works at Portrack on Teesside, Howdon on Tyneside and Hendon on Wearside. Residual quantities of sludges from smaller inland and coastal treatment works would be transferred to the centre by road tanker or to the nearest major centre for shipping.

In making the strategic decision to dry the region's sewage sludge centrally, the company commissioned a study to compare thermal drying and gasification with conventional incineration. The

study was based on a 50 000 tds per annum plant; this is comparable to the first phase of the sludge treatment centre. It was found that drying and gasifying had a number of benefits over incineration. The main findings were as follows.

- Both the capital and whole-life costs were significantly less for thermal drying/gasification.
- Thermal drying/gasification was approximately 4% more efficient in both overall and net electrical generation.
- Drying/gasification was found to be 15% more efficient thermally.
- It has a smaller parasitic electrical load and approximately 1·5 MW more surplus electricity to export.
- Emissions are similar but volumetrically smaller for drying/gasification, therefore the cost of fitting future (legislation driven) pollution control devices would be lower.
- Drying/gasification produces a fuel gas that will replace natural gas. Through the gas turbines this will produce thermal and electrical

Fig. 3. Seaton Carew beach—due to be served by a new sewerage treatment works by the end of 2000

Fig. 4. Projected sewage sludge quantities 1998–2018

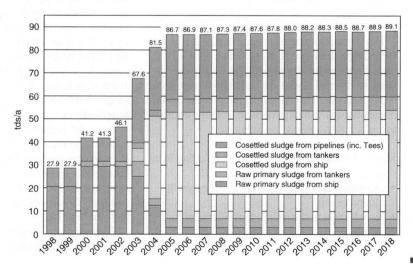

energy, making Bran Sands virtually energy self-sufficient.

- Fossil fuel replacement by drying/gasification has a higher public acceptability in the UK.

Phased development of the sludge treatment centre

Phase 1 of constructing the sludge treatment centre includes utilities sized for subsequent phases. Thus all pipelines and bridges, power supply and distribution cables, gas supply and distribution pipes, chemical and fuel oil storage were completed in 1998.

Phase 1 includes the complete construction of a new jetty, permanent ground-level civil structures such as the effluent sump, cooling tower base and site fencing. It also includes construction of sufficient buildings required to house process plant necessary for treatment of 50 000 tds a year of sludge. Phase 2 will involve extension of these buildings to house the additional process plant required.

The current authorization under integrated pollution control legislation is for a plant with a treatment capacity of 50 000 tds per annum. This plant is the first to be licensed under the legislation.

Scope of the integrated pollution control application

The generic design of the sludge treatment centre was developed using the principle of best available technique not entailing excessive cost (BATNEEC). The facilities to be included within the centre and within the scope of the application will comprise

- sludge reception from ship
- sludge reception from road tanker
- liquid sludge storage
- sludge screening
- sludge dewatering
- sludge drying
- dried sludge storage
- odour control
- combined heat and power generation
- utilities
- infrastructure/buildings.

Sewage sludge drying process

The thermal drying of sewage sludge is a proven technology; there are at present about 110 such plants in operation throughout the EU. A wide range of thermal sludge drying technologies is currently available. The sludge drying process involves the application of heat generated from the combustion of a fuel to achieve the following

- a reduction in moisture content, resulting in an increased calorific value
- a reduction in pathogens (to levels naturally found in soil)
- a reduction in the weight and volume of the sludge.

Fig. 5. Completed regional sludge treatment centre at Bran Sands, Teeside

The company has evaluated, in particular but not solely, two different types of sludge drying technology, which differ in the drying medium, type of drying chamber and drying temperature used. These are as follows.

- Direct drying, where the sludge is brought into intimate and direct contact with a suspending medium such as hot air, combustion gases or superheated steam with temperatures in the range of 110°C for steam and up to 400°C for hot air. The sludge is generally agitated or rotated in the drying chamber to ensure even and complete drying.
- Indirect drying, where the sludge is physically separated from the heat transfer medium, usually saturated steam or thermal oil. The temperature in the drying chamber is in the range of 85°C to 110°C.

Selection process

The company selected the technology supplier (and its particular sewage sludge dewatering and drying technology) for the sludge treatment centre on a competitive tendering basis. A total of 33 enquiries were received from technology suppliers as a result of an advertisement in the *Official Journal of the European Communities*. These 33 companies were sent a qualification questionnaire requesting further information, of which 20 responded to this request. The proposals were systematically evaluated.

From the evaluation, four consortia (technology suppliers/contractors) were chosen to undertake an extensive process design exercise leading to a competitive tender and the eventual award of a design-and-build contract. The design competition provided the opportunity for joint development of the best solution

Following four months of intensive consultation and design, a £20 million contract for the drying

plant was awarded to Andritz AG of Austria utilizing
its direct drying technology.

Regional sludge treatment plant

An overview process diagram for the regional
sludge treatment plant is provided in Fig. 6.

Operation

The plant is designed to operate on a 24-hour,
365-day-a-year programme and individual items of
equipment will be designed for 7000 hours a year
operation. There are no planned periods of complete shutdown, although there will be a planned
rolling maintenance programme. Installation of the
equipment will be phased not only to meet the
envisaged increasing sludge arisings up to the ultimate plant capacity in 2018, but with sufficient
redundancy in-built to allow planned routine maintenance periods to be scheduled such that normal
operations are still maintained. Plant personnel will
operate a four-shift system to ensure 24-hour atten-
dance, 7 days a week.

Liquid sludge handling

Liquid sludge will be transported from Tyneside
and Wearside by sea. In the short term the company's sludge ship will also be used to transfer sludge
from Portrack sewage treatment works. However, in
the near future, Teesside sewage will be re-routed to
the effluent treatment works adjacent to the sludge
treatment centre from which the sludge arisings will
be pumped directly into storage tanks at the centre.
The remaining liquid sludge from the smaller nearby coastal and inland sewage treatment works will
be transported by road tanker to the centre.

Shipped sludge

Sludge arriving by sea (nominally 5% dry solids)
will be carried by the existing company ship with a
capacity of 1500 m³ (Fig. 7). The ship's pumps will
be used to discharge sludge to a jetty side storage
tank, nominal capacity 1500 m³, fitted with a mixer.
The ship will then be washed with water from the
site and washings pumped to the jetty storage tank.
Air from the jetty storage tank will be ducted to an
odour control unit adjacent to the jetty site.

Sludge stored at the jetty will be transferred to the
main site storage tanks along a 2·1 km pipeline. The
pumping arrangement will be configured such that
the rate of pumping may be varied. This arrangement
will minimize the chance of settlement in the pipeline.

Tankered sludge

The remaining liquid sludge from the smaller
nearby coastal and inland sewage treatment works
in the region will be transported by road tanker to
the sludge treatment centre.

Tankers will deposit liquid sludge into a covered
tank. Odorous air will be ducted to the odour treatment unit. Washdown facilities will be provided and
operational procedures will be enforced so that
fugitive odour emissions are eliminated.

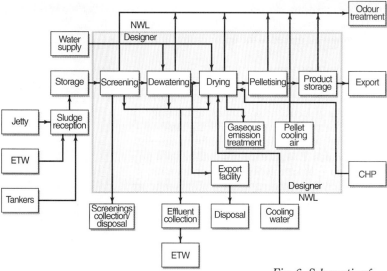

*Fig. 6. Schematic of
sludge treatment cen-
tre process*

Site sludge storage

Tankered sludge, shipped sludge and sludge
from the adjacent effluent treatment works will be
transferred to a mixing tank prior to forwarding for
treatment. To cater for variable delivery rates,
planned and unplanned maintenance, a total sludge
storage capacity of 15 000 m³ will be provided at the
sludge treatment centre. Each storage tank will be
fitted with mixing equipment, in order to prevent
stratification within the storage tank. The tanks will
also be fitted with decanting facilities, which will
allow top water to drain away to the effluent sump.

The liquid sludge will be pumped via a pipework
manifold system to any one of the three process
streams. Indicatively, each stream will be rated to
treat a nominal 16 700 tds of sewage sludge annually,
based on a design throughput of 50 000 tds a year.

Screening of raw sludge

A sludge strainer with a 5 mm perforated sludge
screen will be installed to screen all sludge. The
screenings will be collected in a skip, suitably
designed to reduce odour emissions. All screenings
will be compressed and sent to a suitably licensed
landfill for final disposal (approximately one skip
load a day). The screened sludge discharges from
the strain press and feeds directly into an intermediate tank.

Dewatering

From the intermediate sump, sludge will be
pumped to the belt press dewatering machines,
which will be capable of producing cake at a nominal 29% dry solids.

Optimizing the performance of the 3 m wide belt
presses will be carried out during the commissioning phase. Dewatered cake will discharge directly
to adjacent intermediate cake storage hoppers.
Industrial (raw) water will be used to clean the
belts and the filtrate produced will be collected and
transferred to the effluent sump before transfer to

the effluent treatment works.

The belt presses will be installed within enclosures and odorous air ducted to the odour control unit.

Polyelectrolyte addition

Polyelectrolyte used as a coagulant aid will be added to the liquid sludge prior to dewatering. A common polyelectrolyte preparation and dosing plant will be provided complete with big-bag storage, powder feeder, solution tank, dilution tank and dosing pumps. This package plant will be sized to provide up to 10 kg polyelectrolyte per dry tonne sludge (poly/tds) processed; however, it is anticipated that the belt presses will operate at 4 kg poly/tds.

Cake export

The dewatered cake will discharge into the intermediate hoppers prior to delivery to a drying stream. If they are any major problems with a drying stream, the cake can be diverted to an export facility outside the main building and thus to landfill.

Drying

Cake will be transported from the hoppers to one of the drying streams (Fig. 8). This will be achieved by using screw conveyors.

Dewatered cake will be thermally dried to 90–95% dry solids in the Andritz direct dryer. Sludge is brought into intimate and direct contact with hot air at temperatures up to 400°C. Each drying drum with a 3·2 m ext. dia. and an overall length of 10·6 m (Fig. 9) consists of three concentric cylinders arranged so that the feed material first passes along the innermost cylinder in one direction, then turned 180° to pass through the second cylinder in the reverse direction before turning another 180° to pass through the outermost cylinder. At the end the dried granulate is separated

Fig. 7. Sludge ship MV 'Northumbrian Water' alongside new jetty at Bran Sands

from the circulating air by a bag filter.

The inner walls of the three concentric cylinder drums are fitted with special lifting blades. Together with the circulating air current they transport the sludge forward while it is being dried. Smaller particles arrive at the drum end within a shorter time than large particles, thus ensuring even and complete drying.

Dry product handling

Dry product on leaving the dryer will be immediately cooled in a water-cooled screw conveyor to

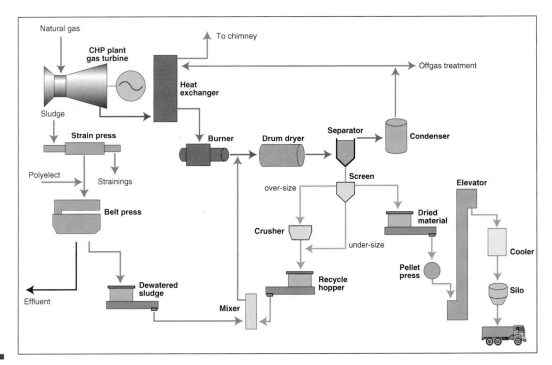

Fig. 8. Schematic of drying line with integrated combined heat and power plant

prevent overheating prior to pelletizing. The dried sludge then feeds into the pelletizing plant. The pelletizing plant will comprise a feed screw, conditioning unit, pellet mill force feeder and pellet mill.

The pelletizing plant is installed in-line prior to the final dry product storage silo in order to combat any perceived problems associated with the handling and storage of low density, fibrous material.

The dry sludge will then be formed into pellets. The final pellet size can be varied by changing the mill die. Trials undertaken to date have successfully pelletized the material in sizes between 4 mm and 12 mm dia.

The temperature of the pellets leaving the mill could exceed 90°C so, prior to final storage, the pellets will be cooled to near ambient temperature. The pellet cooling system will use an air cooler which will both cool and harden the pellets prior to transportation to the storage silos. Extracted air from the product cooler and the conveyor to the storage silos will be treated in the odour control unit.

Product storage

Cooled pellets will be transported to any one of the dry product silos by a pneumatic conveyor system. There will be sufficient storage capacity to hold approximately 5 days production. Silos will be provided with a telescopic discharge system which will allow for filling of lorries for off-site use. The pellet handing system will be totally enclosed. Potential dust will be extracted, filtered and the odourous air routed to the odour control unit. Any dust will be returned to the process for reagglomeration in the dried product.

Condenser

Removal of evaporated water from each dryer system is carried out using a condensing tower. Cooling water is supplied to the condenser heat exchanger from cooling towers located on the site. An effluent bleed is maintained in order to remove any build up of solids in the condenser recycle system. Water removed from the recycle system is replaced by an industrial water feed.

Combined heat and power plant

The treatment centre requires both electrical and heat energy in such a ratio as to enhance the benefits of an on-site combined heat and power installation.

Two gas turbines each nominally rated at 5 MW will generate sufficient electrical power to match the site demand (including the adjacent effluent works) and the exhaust gases will be used as the heat source for the dryers. The power plant will operate on natural gas, with kerosene as a standby fuel. A back-up electrical power supply will be provided from the mains.

The plant will utilize low emissions combustion technology. The full gas turbine train will be located adjacent to the drying building and the majority of the thermal energy required for the drying process will be supplied by the plant.

Hot gas for the direct dryer will be provided from the plant via a hot gas heat exchanger. Top-up and standby heat will be generated by an auxiliary burner providing hot air to the dryers.

Fig. 9. One of the three completed drying lines inside sludge treatment centre

Odour control measures and procedures

In accordance with the company's odour control policy, odour treatment has been designed such that there will be no odour nuisance at either the sludge treatment centre site boundary or the nearest residential property to the site, whichever is most appropriate.

To minimize the likelihood of odour nuisance, the philosophy has been to include specific control measures, such as

- maximum containment of potentially odorous material through the use of totally enclosed equipment and/or ventilated enclosures
- the collection and treatment of all sources of potentially odorous air through a biofilter (main site) or catalytic scrubber (jetty site), with all the connected systems kept under slight negative pressure to minimize the egress of odours
- the collection and treatment of non-condensable waste air arising from each drier stream through a regenerative thermal oxidizer.

Cooling towers

The cooling towers are required to cool water from the dryer condenser heat exchanger and the dried sludge cooler. Cooling towers are of the reduced plume type. The tower works in a cross-flow mode. In this configuration the packing comprises a

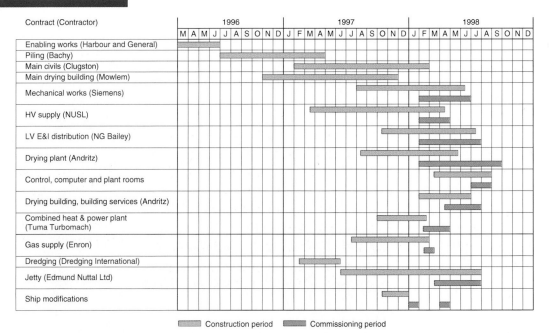

Fig. 10. Construction programme

Contract (Contractor): 1996 / 1997 / 1998 (M A M J J A S O N D J F M A M J J A S O N D J F M A M J J A S O N D)

- Enabling works (Harbour and General)
- Piling (Bachy)
- Main civils (Clugston)
- Main drying building (Mowlem)
- Mechanical works (Siemens)
- HV supply (NUSL)
- LV E&I distribution (NG Bailey)
- Drying plant (Andritz)
- Control, computer and plant rooms
- Drying building, building services (Andritz)
- Combined heat & power plant (Tuma Turbomach)
- Gas supply (Enron)
- Dredging (Dredging International)
- Jetty (Edmund Nuttal Ltd)
- Ship modifications

▭ Construction period ▬ Commissioning period

matrix of parallel horizontal cores with alternate cores operating in wet and dry conditions. The wetted cores carry a film of falling water, and heat is removed into the air stream by evaporation. Air passing through the dry core removes heat from the water by conduction through the packing walls.

The air passing through the wetted cores exits fully saturated and at an increased temperature, but with no change in its absolute humidity. The two air streams are mixed after exiting the cores so the resultant mix is less than fully saturated. When the mixed air exits the cooling tower it will cool down to ambient temperature, but does not reach fully saturated conditions and therefore there is no condensation and hence no visible plume.

Construction

The construction programme is shown in Fig. 10.

Following an extensive site investigation by Exploration Associates, work started on site in 1996 with the enabling works to prepare the site. Table 3 lists all the contractors involved.

The drying plant was handed over on 18 July

Fig. 11. Computer image of gasification column.

1998; first sludge was dried on 28 July 1998. During the following five months the company's operations division, assisted by Andritz and Entec, has undertaken an intense period of familiarization and performance testing. By the end of January 1999 around 90 000 t of wet raw sludge had been successfully dried, with 90% of the dried product—'Biopellets'—going to agriculture. Weather permitting, the company expects to continue delivering a similar percentage of dried product to local farms. Ongoing land reclamation and forestry trials plus several enquiries from potential users indicate a very promising future for the product.

Conclusion

After three and a half years developing Northumbrian Water's sludge strategy and two years of conceptual and detailed design, the regional sludge treatment centre at Bran Sands is a reality. In August 1997 this plant was the first of its type to receive authorization under integrated pollution control legislation, which is an achievement that the company's team is rightly proud of. Under the authorization the company intends that the dried sludge be used as fuel or fuel supplement.

In the near future, gasification of dried sewage sludge (Fig. 11) offers the prospect of generating a fuel gas similar to natural gas. This will be the water industry equivalent of the alchemist's dream—gold from lead—and will be the final chapter of the company's strategy to move from disposal at sea to beneficial reuse. This strategy might not be the most appropriate for other water companies, but NWL believes it is right for the company and the area it serves. The strategy meets all the government's requirements for a sustainable waste management policy and represents the best practicable environmental option to meet the challenge presented by the coastal clean-up.

Green Sea—achieving Blue Flags in Wales

N. R. Lowe, BSc, PhD, FCIWEM

Proc. Instn Civ. Engrs
Civ. Engng,
Millennium Beaches
1999,
132, 53-58

Paper 11715

Written discussion closes
15 November 1999

Keywords: Sewage
treatment & disposal;
environment

Dwr Cymru's bathing water clean-up programme was originally designed to achieve the mandatory standards of the EU bathing water directive using long sea outfalls and minimum treatment levels. However, customer consultation indicated that full treatment, preferably with disinfection, was favoured and this became company policy, later extended to incorporate the achievement of Blue Flag standards. In 1996 Green Sea was launched to deliver further improvements through a joint programme of locating and eliminating the pollution sources. This paper explains how Green Sea is pursuing its aim of achieving both guideline compliance at all designated beaches and 50 Blue Flags in Wales.

Dwr Cymru (Welsh Water) provides sewerage and sewage treatment for the three million people living within its area, which covers most of Wales including the whole Welsh coastline. There are many popular beaches, 68 of which are identified under the 1976 EU bathing water directive (BWD) (Fig. 1). Each year these beaches attract large numbers of visitors from outside Wales, and the income which these tourists generate is an important part of the Welsh economy.

Most of the Welsh population lives near the coast and traditionally much of the sewage has been discharged directly to the sea, so that in 1993 almost half of all sewage received did not undergo biological treatment. At this time regular opinion tracking showed that the company's customers felt strongly about coastal water quality, in particular the lack of sewage treatment at Welsh beaches. There was widespread support for the provision of intensive treatment, especially dis-

infection, instead of relying on marine treatment through the dispersive capacity provided by long sea outfalls.

This support was particularly apparent at

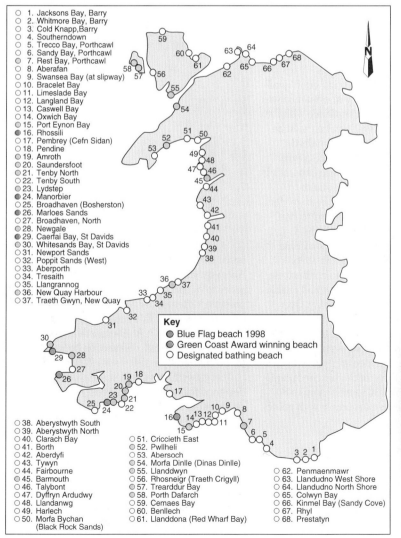

Fig. 1. Location of the bathing waters in Wales identified under the EU Bathing Water Directive

Norman Lowe is chief environmental scientist at Welsh Water

Aberystwyth, where for some years a long sea outfall had been proposed, originally with only preliminary treatment but subsequently incorporating primary settlement. At the end of 1991, however, the National Rivers Authority for the first time indicated that it would accept ultra-violet light (UV) disinfection as part of a permanent sewage treatment solution.

A cost evaluation[1] showed that full treatment and disinfection, discharging to river, was no more expensive than the original scheme; this solution was therefore adopted. A study of the relevant sites in Wales indicated that such a solution was practicable elsewhere, and in early 1993 the company announced its commitment to the eventual provision of full treatment and disinfection at all of its 230 marine sewage works and outfalls. The preferred disinfection process was UV, although other methods would be considered if appropriate.

Shortly afterwards, the company embarked upon a major customer consultation exercise before developing its AMP2 environmental improvement programme for 1995–2000. A questionnaire sent to all customers asked for their views on a range of scenarios relating to expenditure and the extent to which customers would be prepared to pay for them. Although analysis of the results was complex, it was clear that many people were prepared to pay a little more to fund a more rapid and extensive programme for achieving good quality bathing water.

Bathing water quality standards

At first sight, the criterion for judging good quality bathing water was clear. Already in existence was a major programme to achieve compliance with the mandatory standards of the BWD, and it could be argued that full compliance with this standard would be sufficient. But, increasingly, the public and environmental interest groups were expressing their opinion that the BWD guideline standard, some 20 times more stringent, should be regarded as the relevant benchmark. An important reason for this was that the guideline standard was a requirement for the award of the European Blue Flag, won by only two Welsh beaches in 1995.

As a result, the company decided to concentrate its efforts on designing schemes that would enable the achievement of guideline water quality. Over £600 million was already committed to the coastal clean-up, but an additional £38 million to achieve the more stringent standard was then made available from savings made. The declared aim was that by the year 2000 all but three of the beaches identified under the BWD in Wales, as well as up to 80 other beaches, would be able to comply with guideline standards.

Disinfection schemes

Once the 1993 disinfection policy had been declared, first-hand operating experience of disinfection processes was needed so that such a major undertaking could be implemented as smoothly as possible. It was decided to go ahead quickly with an

Table 1. Welsh Water's first UV disinfection plants

Works name	Secondary treatment type	UV type
Criccieth	Biological aerated flooded filter	Wiped, in-channel
Tywyn	Percolating filters	Wiped, in-channel
Aberystwyth	Activated sludge	Unwiped, in-channel
Newgale	Percolating filters	External lamps, pipe

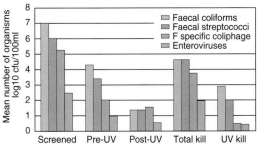

Fig. 2. Microbial kill by sewage treatment and disinfection processes (Aberystwyth)

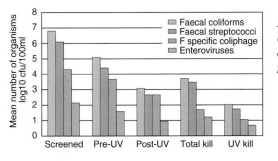

Fig. 3. Microbial kill by sewage treatment and disinfection processes (Criccieth)

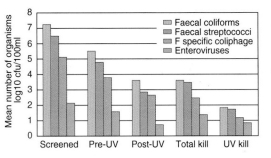

Fig. 4. Microbial kill by sewage treatment and disinfection processes (Tywyn)

initial batch of full-scale UV disinfection projects at Aberystwyth, Criccieth, Tywyn and Newgale. In order to obtain as much comparative information as possible, different disinfection systems were selected, downstream of different sewage treatment processes (Table 1). At Aberporth an alternative process was adopted, disinfection by microfiltration of the chemically-assisted settled effluent.

Aberystwyth was the first such works to be designed following the 1992 cost evaluation. Here a dry weather flow (DWF) of 103 l/s is given preliminary treatment and primary settlement followed by fine-bubble diffused air-activated sludge treatment. Maximum flow to full treatment is 217 l/s and disinfection is by a low-pressure UV system installed in the final effluent channel. There is no automatic sleeve wiping and cleaning is effected by removal and acid-washing. The disinfected effluent is discharged into the river Rheidol 3 km upstream of

*Fig. 5. Swansea
sewage treatment
works*

Aberystwyth harbour.

The three in-channel UV works, Aberystwyth, Criccieth and Tywyn, were monitored for the three years 1995–97 to compare their performance; a summary of the results is shown in Tables 2–4. At Aberystwyth (Fig. 2) a reduction in the mean number of faecal coliforms and faecal streptococci across the works as a whole was greater than $\log_{10}$ 4, and mean numbers in the final effluent were less than 100/100 ml. F. specific coliphage showed a broadly similar pattern, but removal rates of enteroviruses appeared to be considerably less. However, it is considered likely that full recovery of enteroviruses in crude sewage was not achieved because of build up of toxic components following the concentration of the 10 l sample required to achieve a measurable number of plaque-forming units. If the concentration of enteroviruses in crude sewage is in fact under-estimated because of this effect, then removal rates may well be higher than that currently estimated.

Mean removal rates at Criccieth (Fig. 3) and Tywyn (Fig. 4) were generally lower, reflecting less reliable performance, especially in the early part of the trial. At Tywyn, flows rose higher in the channel than had been expected and flooded the control electronics. However, this problem has now been overcome and kills have reliably reached $\log_{10}$ 4. At Criccieth the equipment did not provide the UV intensity expected and was also unreliable, resulting in regular burn-out of the control electronic chokes. It was replaced with a unit similar to that at Tywyn and now produces a comparable performance. At both sites the provision of a single UV unit leads to a risk of total process failure in the event of unit breakdown, and it is now considered preferable to install two units in series.

The results were sufficiently good to encourage the use of UV at a further 10 sites including major works at Tenby, Llanelli, Treborth and Swansea. Swansea (Fig. 5) serves a population of 165 000 and is at present the largest works in Wales to incorporate disinfection. It replaced an old discharge of screened sewage at Mumbles Head, which caused regular BWD failures at several beaches within and around Swansea Bay. Great care has been taken to ensure that the works enhances the local environment. It is fully covered and landscaped and a three-phase dual odour control system has been provided. A maximum flow of 3105 l/s receives grit, grease and screenings removal, following which flows up to 1300 l/s pass forward to full treatment by primary lamella settlement and diffused air activated sludge. Before discharge to Swansea Bay the fully-treated effluent is disinfected by medium-pressure UV lamps installed in the effluent channel.

Green Sea

Increasingly, as coastal sewage treatment was being provided to address BWD standards, it became clear that the company's outfalls were not the only sources of microbiological contamination of coastal water, since even when full sewage treatment was installed compliance with the guideline standards was not always achieved. In 1995, therefore, approaches were made to a wide variety of organizations throughout Wales interested in the quality of the Welsh coastline. It became apparent that there was broad support for an initiative to exploit the opportunity brought about by the improvements in sewage treatment then planned and taking place.

Launched by the Secretary of State for Wales on 1 May 1996, Green Sea is a partnership of organizations committed to working together on a programme of practical action to transform the quality of the sea and coast of Wales. Chaired by the Wales Tourist Board, the Green Sea Forum consists of a wide variety of groups, including Dwr Cymru, local authorities, the Environment Agency Wales (EAW), the Countryside Council for Wales, Keep Wales Tidy, the Marine Conservation Society, Surfers Against Sewage and many more.

The key Green Sea objectives are

- to achieve greater and better co-ordinated practical action to improve dramatically the quality of the coastal environment of Wales
- to publicize this effort and its achievements for the benefits of communities, tourism, the Welsh economy and the public's understanding and enjoyment of the environment
- to create an all-Wales working partnership between the organizations having an interest in improving the environment
- through that partnership to secure a commitment and agreed strategy to work further to improve the coastal environment of Wales
- to achieve 50 Blue Flag award beaches in Wales.

The achievement of 50 European Blue Flag Beaches is a major undertaking. As well as reaching the BWD guideline standards, there are 26 land-based criteria to be met, such as the installation of appropriate access including provision for the disabled, toilets, a water supply and high standards of signs and information for the public. But the potential benefits for Wales are very great. Much of the Welsh economy is directly or indirectly dependent on tourism, and this proportion is even greater in the better-known holiday resort areas such as Pembrokeshire. The Wales Tourist Board regard the cleanliness of the Welsh environment as a strong selling point in their promotion of Wales as a tourist destination, and in addition they report that most tourists visit the coast, whether or not they actually swim in the sea.

Although 68 beaches have now been identified under the BWD in Wales, it may be that not all are

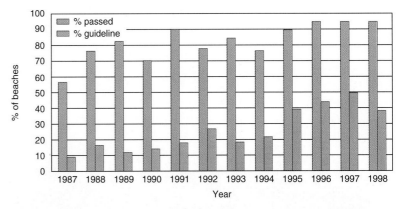

appropriate locations for the Blue Flag as it exists at present. The land-based criteria are relevant to resort beaches, but some would be out of place in a remote, rural site, even if it is well managed. The Green Sea partnership therefore supported the adoption of a 'rural blue flag' based on guideline BWD standards which, while no less stringent than its resort cousin, would contain criteria more suited to the rural location.

In 1998 Keep Wales Tidy, part of the Tidy Britain Group, agreed to hold a trial in Wales of such an award to be known as the Green Coast Award. In conjunction with the Green Sea partnership, draft criteria were developed such that the award would be given to beaches with management plans leading to an appropriate balance between the provision of facilities for beach users and the protection and development of conservation and ecological interests. Nine beaches with guideline standard water quality were selected for this initial trial award, and it is hoped that in 1999 it can be extended to the rest of the UK and perhaps, in time, throughout Europe.

Other sources of pollution

In parallel with the sewage treatment improvements, a programme was developed to find and deal with the other sources of microbiological pollution threatening the achievement of guideline standards. Studies at Jersey[2] had shown that freshwater inputs could be major contributors to the overall microbiological load of bathing water, and it was decided to carry out surveys into catchments in Wales in order to establish the extent to which river inputs were affecting bathing water quality.

The first such study, in 1996, covered the river Nyfer which flows into the sea at the small town of Newport, Pembrokeshire, where one of the two popular beaches (Newport North) is identified under the BWD. Bathing water quality had been variable, with regular failures of the guideline and even sometimes the mandatory standards. Although sewage from Newport is given only preliminary treatment, it is discharged through a long outfall and it was thought unlikely to be the primary cause of the failures. Samples were taken in dry and wet weather during the summer from all the major tributaries and from all significant permanent and intermittent point sources. These showed[3] that

Fig. 6. Bathing water compliance in Wales 1987–1998 showing percentage passing mandatory and guideline standards of EU bathing water directive

*Fig. 7. Dinas Dinlle
beach*

the total input of coliforms and faecal streptococci was much higher in wet weather than the base load in dry weather, which was largely derived from sewage sources. Also, for some parts of the catchment at least, wet weather loads were especially high downstream of improved pasture land, implying a connection with farming practices.

In 1997 a similar study took place on the River Ogwr, which flows through the Bridgend area in south Wales and reaches the sea near Porthcawl. This is larger than the Nyfer and differs in that part of the catchment is urban, with a significant number of combined sewer overflows (CSOs) which operate prematurely. Nevertheless, initial results appeared to indicate that agriculture remains an important source of coliform and streptococcal pollution, although some CSOs were found to be major contributors.

In order to achieve guideline standards reliably it was clearly necessary to find and deal with all significant pollution sources. Consequently, together with the EAW and the relevant local authority, joint sampling programmes were set up to investigate potential pollution around the whole Welsh coast. Priority was given to those beaches such as Aberystwyth, Tywyn, Aberporth, Trearddur Bay and Criccieth where, even though treatment had been installed, guideline standards had either not been achieved or were threatened. The studies concentrated on point sources and made much use of local knowledge to pin-point the areas where pollution was likely to originate. To date, they have been successful in locating pollution from CSOs, private discharges and foul sewage cross-connections to surface water drains, all of which can then be tackled to bring about the necessary water quality improvements.

The EAW is now working with farmers to modify agricultural practices to reduce microbiological pollution, particularly in areas identified as potentially contributing to guideline failures. This process is assisted by the fact that the National Farmers Union in Wales is a member of Green Sea and sup-

*Fig. 8. Port Eynon
beach*

port its aims. In view of the fact that incidents of high microbial numbers were typically associated with high river flows, it is believed that heavy rainfall and the associated surface run-off are implicated. Work is therefore concentrating on farm management of slurry and farmyard manure to minimize the likelihood of such material being washed into streams in wet weather.

Although work is still under way on both sewage treatment and other pollution sources, water quality has improved markedly (Fig. 6). Between 1987 and 1997, bathing water compliance with BWD mandatory standards rose from 57% (27 out of 47) to 94% (60 out of 64), and with the guideline standards from 9% (4 out of 47) to 48% (31 out of 64). However, in 1998 the wet weather during the bathing season affected the results, and although compliance with mandatory standards was maintained at 94%, guideline compliance fell to only 37% (25 out of 68). Beaches which did reach guideline standard included Llanddwyn on Anglesey, Dinas Dinlle on the Lleyn Peninsular (Fig. 7), Port Eynon in the Gower (Fig. 8), and Tenby North in south Pembrokeshire (Fig. 9).

In 1998 only one of the mandatory failures, Swansea Bay, was associated with pollution from a sewage outfall, in this case from the Mumbles outfall serving Swansea. At the other locations, Poppit Sands, Criccieth and Abersoch, full treatment is installed at the local sewage discharges, and the failures are attributed to other sources, especially those aggravated by heavy rain such as storm discharges and agricultural run-off. It was particularly disappointing that Criccieth failed mandatory standards in view of the UV disinfection at the treatment works, but it did highlight the importance of ensuring that all polluting discharges are identified and dealt with before good water quality can be assured.

Conclusion

Since 1987, when the first large-scale identification of BWD bathing waters was made in the UK, major investments have been made to achieve compliance, originally with the mandatory standards. But in the meantime, public perception of the importance of bathing water quality has heightened and at the same time increasing emphasis is being placed on the importance of the guideline BWD standards and the award of the European Blue Flag.

The more stringent standards inevitably necessitate additional treatment at some locations, but in Wales the additional £38 million contributed by Dwr Cymru for this was relatively small in comparison with over £600 million committed to coastal waste water improvements in the five years 1995–2000. However, the improvements have revealed other sources of microbiological pollution affecting guideline compliance. Foremost are land run-off and the remaining CSOs, and it is especially important that the programme to deal with all unsatisfactory CSOs progresses quickly.

Against these extra costs should be set the benefits

Fig. 9. Tenby North beach

of guideline compliance. From the water company's point of view, customer satisfaction is of course important. But there are other, more tangible, benefits. A large number of Blue Flags awarded each year throughout Wales will build a perception of a clean coastal environment. It is expected both that residents and potential visitors will be more likely to spend holidays in Wales, thereby contributing to the economy, and that industry will be more likely to stay in or relocate to Wales as a result of this positive perception.

Much of the attraction of the Welsh coastline stems from the existence of many smaller, relatively remote beaches, at which the infrastructure associated with the Blue Flag is not appropriate. Recognition of these by the Green Coast award, based on the guideline standard, will be a further step towards reaping the benefits of excellent water quality along the whole coast.

Dwr Cymru's investment in coastal sewage treatment is very large, and it is vital that it achieves value for customers' money. Through co-operation with Green Sea partners, the opportunity exists to achieve guideline BWD standards at a large number of beaches. In turn, this will enable the award of many more Blue Flags and equivalent awards, emphasizing the high quality of Welsh beaches and boosting tourism and the economy of Wales.

References

1. GIBSON M. T. Reasons for choosing disinfection for a coastal town. In *Disinfection of Urban Wastewater*. Water Research Centre, Swindon, UK, 1992.
2. WYER M.D., KAY D., JACKSON G.F., DAWSON H.M., YEO J. and TANGUY L. (1995) Indicator organism sources and coastal water quality: a catchment study on the Island of Jersey. *Journal of Applied Bacteriology*, **78**, 1992, 290-296
3. WYER M.D., CROWTHER J. and KAY D. (1997) *Faecal Indicator Organism Sources and Budgets for the Nyfer Catchment, Pembrokeshire—A Report To Dwr Cymru And The Environment Agency.*

Yorkshire Water's Coastcare

P. Langley, CIWEM, R. Stringer, CEng, MIStructE and G. Lang, MICE

Proc. Instn Civ. Engrs
Civ. Engng,
Millennium Beaches
1999,
132, 59-64

Paper 11716

Written discussion closes
15 November 1999

Keywords: Sewage
treatment & disposal

Coastcare is Yorkshire Water's £120 million coastal waste-water treatment improvement programme. It involves installing modern treatment plants to serve the popular east coast resorts of Scarborough, Filey, Whitby, Bridlington, Hornsea, Staithes, Robin Hood's Bay, Sandsend and Runswick Bay. Developed in close consultation with the local communities, the programme takes account of potential future changes in legislation. It thus moves away from the old solution of long sea outfalls towards inland treatment technologies, including UV disinfection.

Coastcare is Yorkshire Water's £120 million coastal sewage treatment improvement programme (Fig. 1). It includes plans to install modern waste-water treatment plants, including the latest ultra-violet treatment, at the popular east coast resorts of Scarborough, Filey, Whitby and Bridlington. Together with planned investment at Hornsea, Staithes, Robin Hood's Bay, Sandsend and Runswick Bay (Fig. 2), the work will greatly improve the amenity value of the Yorkshire coastline for both local people and holidaymakers.

Much of the sewerage system in the Yorkshire region's coastal towns was built by the Victorians in the early 1900s. As the tourist industry grew over the years, waste water in these towns was usually directed to short pipes which discharged untreated waste water into the sea. This meant that visible waste-water solids were sometimes washed back onto our beaches and the coastline.

The Yorkshire coastline has 22 EU designated bathing beaches and 12 significant discharges. All these discharges are on or near bathing beaches, and so the requirements of the EU bathing water directive (BWD) and urban waste-waters treatment directive overlap to a considerable extent—the solutions to both being inter-related. In 1994 the UK government declared all the receiving waters pertaining to our coastal investments as high natural dispersion areas (HNDA) under the terms of the urban waste-waters treatment directive. This meant that each of the 12 discharges required preliminary and, in some cases, primary treatment to comply with the directive.

A few of the discharges, notably Scarborough, Bridlington, Flamborough and Withernsea, had long sea outfalls installed in the early 1990s to comply with the BWD, the first two passing screened sewage and the second two primary treated sewage from works built by local authorities. Apart from this, little work had been carried out since the short outfalls were installed in the early part of the century. In addition, combined storm overflows (CSOs) discharging to low water or inland streams, which in turn flow over the beaches, made matters worse.

Table 1 shows compliance with the bathing water directive since 1990 for a total of 22 designated beaches. Failures are in part due to the discharge of untreated sewage via both old short sea outfalls and CSOs and in part to private discharges, for example private caravan sites, direct industrial discharges and run off from agricultural land.

During 1996, as a result of management changes, the water company reviewed all its policies with a particular emphasis on consultation with local opinion-formers, including MPs, councils at all levels and representatives of pressure groups. It was agreed that coastal sewerage issues were second only to reliable water supply and leakage in the public's mind. From this review the Coastcare policy was born.

Fig. 1. Yorkshire Water's Coastcare schemes

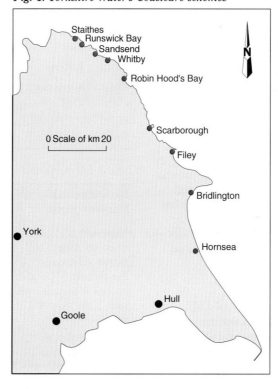

Peter Langley is capital investment manager at Yorkshire Water Services

Roy Stringer is senior project manager at Babcock Water Engineering Ltd

Graham Lang is project manager at Montgomery Watson

Coastcare

The company agreed that the sewage from the four larger towns of Whitby, Scarborough, Filey and Bridlington was to be given full primary, secondary and tertiary (UV) treatment prior to discharge. At the five smaller towns and villages of Staithes, Runswick Bay, Sandsend, Robin Hood's Bay and Hornsea, appropriate solutions were to be brought forward from the urban waste-waters directive compliance date of December 2005 to December 2000. In practice these will also be subject to secondary and UV treatment, either as separate waste-water treatment works or by pumping to one of the larger treatment works created as part of this programme.

There are four principal reasons for instigating the Coastcare policy. First, it is a result of listening to customers, their concerns at current practices and their desire to see Blue Flag beaches and an enhanced tourist industry. It is interesting to note that in Yorkshire, as elsewhere in the UK, diffuse and private discharges will continue to affect bathing waters, and in some cases may result in non-compliance with the current standards irrespective of the investments made by the water company.

Second, it is in anticipation of future legislation, particularly following the Tenby judicial review which has since resulted in the Environment Agency's new coastal water consenting (virus) policy, the revocation of HNDA status and potential revisions to the BWD which are being discussed in the EU. These discussions may result in a 50 faecal streptococci guideline standard being applied to scheme designs compared to the 2000 faecal coliform mandatory standard currently in force.

Third, it is good environmental practice. One of the aims of the urban waste-water treatment directive is to reduce the level of nutrients entering the North Sea and to minimize public health risks and litter problems for residents and tourists in bathing areas.

Finally, it makes good business sense. Long sea outfalls, although adequate for the current BWD compliance, will not comply with proposed future requirements or the latest EA virus policy. New long sea outfalls are therefore likely to be out of date even in the short term. To construct works with a higher level of treatment now will result in optimal solutions for the future. In addition, installing full treatment now increases the potential for contractors to develop efficient designs and reduces the total administrative overheads by having a single scheme and contract. To return to carry out further extensions at regular intervals as legislation advances would result in sub-optimal designs and duplicated effort.

Partnership

The development and implementation of Coastcare has increasingly led to working partnerships with local authorities, contractors and public and business representatives, which are essential if the company is to achieve the required benefits for

Fig. 2. Runswick Bay looking north west (courtesy Peter Smith Photography, Malton)

Table 1. Number of beaches passing since 1990

Year	Mandatory standard	Guideline standard
1990	17	0
1991	17	2
1992	18	2
1993	17	4
1994	17	3
1995	13	7
1996	18	2
1997	18	1
1998	19	2

Table 2. List of principal consultants and contractors (Bridlington)

Project element	Project management consultant	Contractor
Contract 1 (headworks)	Babcock Water Engineering Ltd	Birse Construction Ltd
Contract 2 (waste-water treatment works)	Babcock Water Engineering Ltd	Birse Construction Ltd
Gypsey Race sewerage	Babcock Water Engineering Ltd	Miller Civil Eng. Ltd
North Bridlington CSOs	Babcock Water Engineering Ltd	Morrison Construction

Table 3. List of principal consultants and contractors (Staithes and Runswick Bay)

Consultant/contractor	Services provided
Montgomery Watson	Project management and engineering services, preparation of environmental statement
Wilbraham & Co.	Planning services and advice
Clive Brook Associates	Planning services and advice
Fawcett and Fawcett	Landscape and visual services and advice
Carl Bro Group	Highways and traffic services and advice
Northern Archaeological Associates	Archaeological services and advice
Webb Seegar Moorhouse	Architectural services
SGS Environment	Environmental services and advice
Chris Blandford Associates	Environmental (noise) services and advice
Bullen	Marine modelling services and advice
Taylor Woodrow Construction	D&C contractor

the resorts, the environment, tourism, beaches and bathing waters.

The following are two examples of the development of Coastcare schemes, one at the larger coastal resort of Bridlington (summer population equivalent of 80 000) and one at the small fishing villages of Staithes and Runswick Bay (combined summer population equivalent of 3500).

Bridlington

Bridlington is a popular holiday resort on the east Yorkshire coast. The sea and the bathing beaches are key elements in its existence as a tourist centre, and their cleanliness is central to the town's leisure industry.

The town has a permanent population of approximately 34 000, which rises to 55 000 in the summer months, plus trade discharges which combine to give a peak population equivalent of 80 000. Focal points for these visitors include the beaches and the harbour.

Babcock Water Engineering Ltd was commissioned to develop a strategy to resolve deficiencies within the Bridlington drainage area. By looking at the whole catchment, the consultancy was able to develop individual schemes which, when completed, provided a cost-efficient solution to all of the problems.

The Bridlington sewerage system consists of a dendritic network of gravity sewers which converge at the long sea outfall headworks (Fig. 3). The long sea outfall has a formula A (1200 l/s) capacity, and all dry weather flows receive fine screening before discharge. Under storm conditions all flows are pumped down the outfall.

With the system there are a series of unsatisfactory CSOs that discharge storm sewage to the bathing waters and an inland water course (Gypsey Race), which runs into the harbour. Sewage debris was frequently deposited on the beaches after storm events. The impact of the CSOs was investigated through a sewerage study using a hydraulic model of the whole catchment. Schemes were developed for the Gypsey Race CSOs, North Bridlington CSOs and the long sea outfall head-

Fig. 3. (above). South Sands beach in Bridlington showing long sea outfall and headworks on Promenade

Fig. 4. Location of Bridlington's new inland waste-water treatment works

works (incorporating the new waste-water treatment works flow transfer modifications).

Gypsey Race sewerage

Four CSOs initially discharged to Gypsey Race and then to the harbour. The EA stated that the existing consents to discharge were to be revised, involving the provision of storage based on the Scottish Development Department design approach.

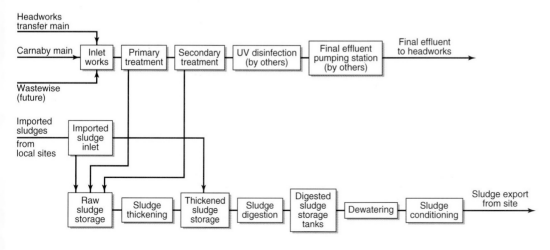

Fig. 5. Process diagram for the Bridlington scheme

Using models based on Wallrus (subsequently Hydro Works), a cost–benefit approach was agreed with the EA to determine the balance between storage required and acceptable CSO discharge. Storm screening was also provided. An off-line detention tank was to be constructed for one of the CSOs and the other two were to be reconstructed with on-line storage being mobilized with a combination of ovoid pipes and pre-cast culverts. The need to construct the whole scheme within one winter period was identified at an early stage to avoid disruption during the peak tourist season.

During contract preparation, problems were identified with one of the sites. In order to mitigate any potential delays, the scheme was split into two construction packages. The company was investigating the use of the *Engineering and Construction Contract* (ECC) form of contract and, in light of this split, chose to test the different contract options available. The first phase involved the construction of the off-line detention tank. The contract was let as a contractor-designed package using ECC option A (activity schedule), with Miller Civil Engineering Ltd being the successful tenderer.

The potential problems of co-ordinating two contractors on two independent sites were removed when the first-phase contractor successfully negotiated for the second phase of the works. This was let as a consultant-designed package using option B (bill of quantities) of the ECC. The contractor used the same site team to run both contracts, which also reduced overall costs.

On the initial scheme budget of £5 million, savings totalling £2 million have been identified so far.

North Bridlington CSOs

This £2 million scheme involved three CSOs in the north of the town. They discharged unscreened storm sewage to sea through two short outfall pipes. Initially, Wallrus hydraulic modelling was used to assess options for improving their performance to satisfy the requirements of the EA. Storage had to be provided to reduce the spill frequency to three per bathing season.

Using results obtained from WRC's Simpol spreadsheet model and 25-year rainfall data, the consultant demonstrated that the spill frequency could be doubled without compromising the status of the bathing beach. This resulted in a total storage reduction in excess of 50%.

The scheme subsequently designed included an on-line detention tank (830 m³), on-line storage (550 m³) and the extension of a small pumping station wet well. The scheme was let as a single package under the ECC form (option A) based on the consultant's design. Morrison Construction was the successful tenderer.

Bridlington waste-water treatment works and flow transfer

The proposed scheme is to provide green-field sewage treatment facilities at a waste-water treatment works located to the south east of Bridlington at Flashdales (Fig. 4). Flow is to be transferred from a modified headworks via a new transfer pumping main to the plant for treatment before being returned to the headworks for discharge through the existing sea outfall. The waste-water treatment works would also become a sludge treatment centre to treat sludge both from the works and from outlying villages (Fig. 5).

The existing headworks will be extended and modified and transfer mains will be provided to

- ensure that increased flow can be passed to the long sea outfall to limit the impact of the storm overflows on the receiving waters
- allow all flows up to a 1-in-5-year storm event to receive screening down to 6 mm
- provide flow transfer pumps and mains to pump flow to the proposed treatment works and provide facility to transfer and receive treated returned flows (flow to full treatment 24 869 m³/day)
- provide chemical dosing for septicity control.
- provide odour control.

The proposals for the headworks are required to interface with extensive refurbishment works which are currently being undertaken by East Riding of Yorkshire Council to the existing promenade.

Full treatment facilities will be provide at the Flashdales site, including

- inlet works with fine screens (6 mm) and grit removal facilities
- primary settlement
- activated sludge secondary treatment (fine bubble)
- final settlement
- sludge digestion and conditioning
- ultra violet disinfection
- odour control.

Planning. The site of the proposed treatment works was purchased by Yorkshire Water in March 1994, and was subsequently confirmed as the most appropriate site following a comprehensive site selection process led by planning lawyer Wilbraham & Co. of Leeds. Planning permission for a primary treatment works and associated sludge treatment plant on the site, refurbishment works at the headworks, together with a pipeline corridor for the transfer mains from and to the headworks, was approved in August 1996.

Following the approval of the Coastcare policy, which provides full treatment and UV disinfection facilities at Bridlington, a revised planning application and environmental impact study was submitted to the planning authorities, planning approval for the full scheme being gained in December 1997.

Contract strategy. To establish the scope of the works a detailed feasibility design was produced.

The design was refined during value engineering and risk management studies. It was also identified that the IChemE (*Red Book*) form of contract was, in this instance, the most appropriate form. Two contracts were to be let, one for the headworks and transfer pipeline and the other for the waste-water treatment works.

Programme. In September 1997, the contract for the headworks and flow transfer pipeline was let to Birse Construction. The contract has been sub divided again into two discrete phases. The first phase has been undertaken during winter 1997/8 and is for the construction of all civil works, while the second phase is being undertaken during 1999 and is for the installation and commissioning of all mechanical and electrical works. The second phase coincides with the commissioning of the waste-water treatment works.

The contract for the waste-water treatment works was let in May 1998, also to Birse Construction, with take over programmed for December 1999. It is the target of the programme to have full treatment available for Bridlington for the bathing season of the year 2000.

Works to date. Construction work commenced on site in October 1997 for phase 1 of contract 1 and was handed over to East Riding of Yorkshire Council in June 1998 for integration into its promenade redevelopment scheme ; the pipeline corridor was cleared for the summer season. Construction of the waste-water treatment works is on programme for completion in December 1999.

Staithes & Runswick Bay

Staithes (Fig. 6) and Runswick Bay are small communities located on the rugged coastline to the north of Whitby in north Yorkshire. Situated within the North York Moors National Park, this area is a focal point for tourists and is important for its landscape, ecological and historical interest. The constraints placed on development in such an area, together with some of the particular physical difficulties of the locality, present an interesting challenge to the provision of new sewage treatment facilities under Coastcare.

The two villages, located approximately 4 km apart, are surrounded on three sides by steeply rising ground, with elevations up to 100 m AOD, and on the fourth side are open to the unpredictable nature of the North Sea. Access is gained by minor roads with gradients up to 1 in 4, and much of the lower parts of the two villages are served by narrow passageways which provide for pedestrian access only. The steepness of the ground and the effects of coastal erosion have led to very unstable ground conditions in a number of places.

Each village is presently served by a combined sewerage system which discharges untreated sewage into the sea via a short outfall in close proximity to an EC bathing water. Under Coastcare, the

Fig.6. Staithes harbour (courtesy Peter Smith Photography, Malton)

water company has committed to provide a scheme for both communities to meet the requirements of the urban waste-waters treatment regulations ahead of the 2005 deadline and to eliminate the risk of bathing water failures due to water company discharges.

The peak summer population equivalent (pe) of the two villages is about 3500 and the appropriate urban waste-waters treatment standard can be achieved by the provision of fine screening and extended outfalls. The feasibility study, carried out by Montgomery Watson, considered this 'base obligation' together with a number of other separate and combined solutions to serve the two villages

- screening at a new inland headworks location and discharge via new long sea outfall(s)
- secondary treatment at a new inland works and discharge via existing outfall(s)
- transfer of flows for treatment and/or discharge at a remote catchment (e.g. Whitby).

The collection point for each of the existing systems is on the foreshore, below high tide levels. The option of providing treatment facilities in these foreshore locations was quickly discounted, primarily due to the limited land available and the restricted access. The provision of a number of new pumping stations to intercept flows for transfer inland were thus investigated.

At each location, stormwater detention tanks and storm overflows are to be provided to attenuate and/or discharge excessive storm flows. The pass-forward flows for either treatment or onward transfer are, therefore, relatively low (3DWF = 19 l/s), but the pumping heads are considerable, typically in the range 60 m to 90 m. This combination of low flow and high head presents particular difficulties for sewage pumping and a range of options has been considered, including staged pumping, coupled centrifugal and progressive cavity pumps. The low night-time flows combined with the relatively long transfer distances also gives rise to the potential for septicity within the rising mains; careful consideration has, therefore, been given to the provision of anti-septicity dosing and odour control equipment.

The option to transfer flows to a remote catchment was eventually ruled out on technical grounds and an inland site was thus required for the provision of either a screening plant or a full treatment facility. At an early stage in the feasibility study it was recognized that, whatever the option finally chosen, the sensitivity of the area and the emotive nature of sewage treatment would mean that obtaining planning approvals would be a critical activity. A dedicated planning team was therefore established and a comprehensive site selection study, led by the planning lawyer, was undertaken. The consultant provided the engineering input to the study and other specialist members of the team included planning consultants, landscape architects, highways and traffic engineers, archaeologists, ecologists and architects.

The site selection study followed a common methodology that Yorkshire Water had developed for all the new waste-water treatment works required within the Coastcare programme. The methodology involves the application of a three-stage sieve process to an area of search defined by a number of initial assumptions. At each of the three stages a range of planning, environmental, technical and financial criteria is applied in increasing levels of detail to reducing areas of land until a preferred site is identified. Once this exercise had been completed for Staithes and Runswick Bay, it became apparent that a combined works to serve both communities was preferred; the final appraisal of scheme options against engineering, operational and economic criteria was then based around this site.

Following completion of the feasibility study, the proposed scheme includes

- the interception of flows by a system of up to three pumping stations
- pressure and gravity transfer pipelines
- stormwater detention tanks and screened storm water outfalls
- a new waste-water treatment works to provide secondary treatment and ultraviolet disinfection
- final effluent discharge via the existing outfall at Staithes.

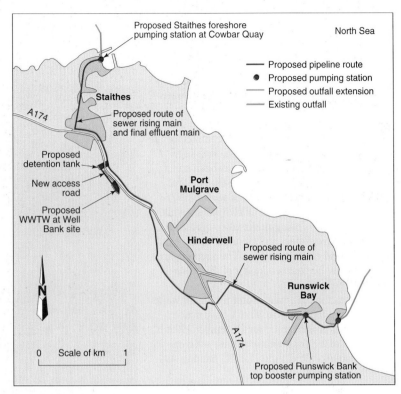

Fig. 7. Layout of Staithes and Runswick Bay scheme

Figure 7 shows the layout of the proposed scheme.

The feasibility study included the development of sewer and marine dispersion models to assist in the identification of the optimum solution for the detention, treatment and discharge of final effluent and stormwater. Output from these models is to be used to support discharge consent applications to the Environment Agency.

A detailed environmental assessment was undertaken for these proposals and a formal environmental statement was submitted with a planning application to North York Moors National Park Authority. An IChemE *Red Book* Design and Construct Contract has been awarded to Taylor Woodrow Construct Ltd and, subject to planning approval, it is proposed to commence construction in September 1999 to achieve beneficial completion before the end of December 2000.

Conclusion

Yorkshire Water's Coastcare policy has been developed in consultation with, and with the full support of, the local communities. It is a forward-looking policy which takes account of potential future changes in legislation and aims to provide optimal solutions bearing future needs in mind. As a result, it moves away from the old solution of long sea outfalls and to solutions based on inland treatment technologies including UV disinfection. In doing so it fully supports the tourist industry and the demands of our customers, produces optimal design solutions and prepares the company for future changes in legislation.